Rethinking Expertise

Rethinking Expertise

HARRY COLLINS AND ROBERT EVANS

The University of Chicago Press
Chicago and London

Harry Collins is distinguished research professor of sociology and director of the Center for the Study of Knowledge, Expertise, and Science at Cardiff University. He is the author of, most recently, *Gravity's Shadow: The Search for Gravitational Waves* and, with Trevor Pinch, *Dr. Golem: How to Think about Medicine*, both published by the University of Chicago Press. **Robert Evans** is senior lecturer in sociology at the Cardiff School of Social Sciences.

The University of Chicago Press, Chicago 60637
The University of Chicago Press, Ltd., London
© 2007 by The University of Chicago
All rights reserved. Published 2007
Printed in the United States of America

16 15 14 13 12 11 10 09 08 07   1 2 3 4 5

ISBN-13: 978-0-226-11360-9 (cloth)
ISBN-10: 0-226-11360-4 (cloth)

Library of Congress Cataloging-in-Publication Data

Collins, H. M. (Harry M.), 1943–
    Rethinking expertise / Harry Collins and Robert Evans.
        p.   cm.
    Includes bibliographical references and index.
    ISBN-13: 978-0-226-11360-9 (cloth : alk. paper)
    ISBN-10: 0-226-11360-4 (cloth : alk. paper)
    1. Knowledge, Sociology of.   2. Expertise.   3. Science—Social aspects.
I. Evans, Robert, 1968–   II. Title.
HM651.C64 2007
158.1—dc22

                                                        2007022671

⊗ The paper used in this publication meets the minimum requirements of the American National Standard for Information Sciences—Permanence of Paper for Printed Library Materials, ANSI Z39.48-1992.

To everything there is a season . . . a time to break down, and a time to build up.

ECCLESIASTES

CONTENTS

ILLUSTRATIONS

PREFACE AND ACKNOWLEDGMENTS

This book is the outcome of a long and difficult journey which began in earnest in the mid-1990s. We thank a Cardiff University group who in those early days helped to put together a research grant bid on expertise in the public domain. We did not get the grant but we worked out a lot of ideas with the team. The loyal members of the weekly KES group seminar have helped us refine the ideas in the intervening years. We thank Tammy Boyce, Simon Cole, Mike Gorman, and the late Jon Murdoch for their work on previous versions of this project and for their endless tolerance. We also thank those who volunteered to read early drafts for us and offered us support as academic colleagues; Martin Kusch and Evan Selinger are but two of a larger number. We are grateful to the wonderfully positive and engaged audiences at the many lively conferences, workshops, and seminars where we have presented aspects of this thesis, and we thank those who have already chosen to use these ideas in their work. We even thank those colleagues whose attempts to stop our work from being published, or even referenced, make novels of academic life seem dull; they reassured us that whatever we were doing it was not run-of-the-mill. And we especially thank those of our referees who grasped the spirit of our project before criticizing—we learned much from them. The University of Chicago Press has been wonderful throughout. Joel Score was assiduous and perceptive in helping us polish the final text. We owe a special debt to our commissioning editor, Catherine Rice, for her courage. During the difficult times, Chicago, Catherine, and, at the final stage, Christie Henry, did everything right.

Why Expertise?

Science, if it can deliver truth, cannot deliver it at the speed of politics. The idea that science would one day be able to solve all problems by the application of logic and experiment began to fail at the beginning of the twentieth century. Quantum theory, Gödel's proof, the turning in on themselves of philosophies such as logical positivism and, more recently, the rediscovery of chaos, have shown that, as in the nightmare, the train of a perfect science is always leaving the station just as you get there. And those are just the "internal" problems.

In the middle of the century Thomas Kuhn's famous book *The Structure of Scientific Revolutions* was seen by some to replace the idea of orderly progress in science with mob psychology. Subsequently a series of carefully documented studies of the day-to-day unfolding of scientific life, especially scientific controversies, showed that the "canonical model" of science did not coincide with the practice itself. The latter part of the century saw a growing public distrust in science springing from the highly visible failures of major technologies and the disasters associated with them, from the manifest politicization of debates about scientific progress in fields related to biology, and from the evermore evident lack of exact understanding by scientists of the legacy of fission power and the risks posed by new agricultural practices. Political movements associated with environmentalism and animal rights have bolstered the distrust in science and technology, while the detailed studies of science which emerged from the social sciences have been swamped by the much more widespread movement known as "postmodernism" springing from the literary criticism.

Taken together, these tendencies seem to have given rise to a *Weltanschauung* in which we no longer understand how to balance science and technology against general opinion. In today's world the scales upon which science is weighed sometimes tip to the point where ordinary people are said to have a more profound grasp of technology than do scientists. Our loss of confidence in experts and expertise seems poised to usher in an age of technological populism.

We need a way to speak and think about science and technology that is not hostage to science's newfound epistemological weaknesses and short-term political impotence. In this book we move from evaluating science as a provider of truth to analyzing the meaning of the expertise upon which the practice of science and technology rests. Perhaps this will help us understand how it can be that though science and technology do not touch the divine they are still the best way to distill human experience of an uncertain world. The underlying assumption of this analysis is that, other things being equal, we ought to prefer the judgments of those who "know what they are talking about." This does not mean that correct judgments are always made by those who know what they are talking about. On the contrary, a good part of the time experts' judgments turn out to be wrong. In some cases we know in advance that experts' judgments are likely to be wrong because they have been so wrong in the past. The assumption means simply that in spite of the fallibility of those who know what they are talking about, their advice is likely to be no worse, and may be better, than those who do not know what they are talking about. Of course, this commonsense notion rests on a raft of ambiguities, for example, the question of exactly what it is that is being talked about in any particular case. We will return to these problems after the technical discussion. First we need to work out what it means to know what you are talking about.

To understand what it is to know or not know what you are talking about we need a new sociology of expertise. There is already a sociology of the acquisition of expert status which shows that coming to be called an expert may have little to do with the possession of real and substantive expertise. This book is not intended to add to the analysis of the process by which experts acquire expert status; it is meant to *make a contribution* to the process. It is meant to increase the chance that the process of coming to be called an expert will have more to do with the possession of real and substantive expertise. To treat expertise as real and substantive is to treat it as something other than *relational*. Relational approaches take expertise to be a matter of experts' relations with others. The notion that expertise is only an "attribution"—the, often retrospective, assignment of a label—is an example of a relational theory. The realist approach adopted here is different. It starts from the view that expertise is the real and substantive possession of groups of experts and that individuals acquire real and substantive expertise through their membership of those groups. Acquiring expertise is, therefore, a social process—a matter of socialization into the practices of an expert group—and expertise can be lost if time is spent away from the group. Acquiring expertise is, however, more than attribution by a social group even though acquiring it is a social process; socialization takes time and effort on the part of the putative expert. In the case of relational theories, on the other hand, all the work is done by the attributors. Under our treatment, then, individuals may or may not possess expertise independently of whether others think they possess expertise.

To give a simple example, in France everyone can speak French, "even the little children," and it is not thought of as an expertise. On the other hand, in Britain a person who is fluent in French is thought of as an expert and can, for example, command a salary as a translator or teacher. It's the opposite way round in France, where it is speaking English that counts as the useful expertise. In a purely relational theory the expertise involved in speaking French and English is no more nor less than that attributed to speakers of the languages in their respective countries. In a realist/substantive analysis, on the other hand, the degree of expertise in speaking a language remains the same in whichever country the language is spoken.

Equating the ubiquity of an expertise with the absence of an expertise has been responsible for some serious mistakes. For example, before attempts were made to make natural-language-handling computers, the deep difficulties of speaking a language were not understood. During that period a *mistake* was being made in treating native language-speaking as easy, a mistake that revealed itself soon after attempts were made to construct machines to do it. Relational or attributional theories have no way of discussing whether language-handling computers really are capable of performing as advertised or why they might not be. Under relational theories the success of language-handling computers is a matter of their degree of acceptance by their users; under a realist/substantive theory differences between computers' and humans' language handling would remain salient even if computers turned out to be so useful in practice in spite of their poor language handling that, in day-to-day life, people ceased to notice the difference between their performance and that of humans.

In chapters 1 and 2 we present ways to think about the substance of expertise organized in a table which, "to get the ball rolling," we call the "Periodic Table of Expertises." The special claim of our classification is that it has an internal structure that makes it more than a hierarchy. It is theory-laden, the idea of tacit knowledge being the chief organizing principle.[1] One of the new distinctions in "The Table," that between inter-actional and contributory expertise, will be analyzed in depth and even empirically investigated. In due course we hope that other distinctions found in the table might be similarly treated.[2] If the approach is success-ful, it will make it seem odd and crude that anyone would ever have spo-ken simply of the rights of "experts" on the one hand and "laypersons" on the other without taking into account the many different ways of being an expert, the distribution of differing expertises among different groups, and the relations between these groups.

A word of warning: however much we analyze expertise, we are not going to be able to develop a complete solution to the problem of the relationship between experts and specialists on the one hand, and lead-ers, generalists, and democracy on the other. The argument is at least as old the Greek city state. Plato's Philosopher Kings may have been experts, but the question remained: "Who shall guard the guardians?" The unre-solved tension between expertise and democracy reappears wherever experts and specialists have found themselves engaged with other groups, not least in the contemporary debate about science and the citizen. The debate occurs within science itself as experiments become so large that professionals such as managers, accountants, and engineers have to be brought in to run them. Who ran the Manhattan Project? Was it Robert Oppenheimer or General Groves? Did even Oppenheimer have a com-plete grasp of all the science he was supposed to be leading? As the key phrase from old debates about the relationship between scientists and the British Civil Service put it, should scientists be "on top" or "on tap"? Should the Civil Service itself be run by generalist administrators, or should specialists play a more important role? Did Senator Proxmire have any right in the 1970s and 1980s to claim that he could spot when a sci-entific project funded by the National Science Foundation was a sham—a "golden fleece," as he called it—or was he an ignoramus who should have left scientific judgments to the tacit instincts of scientists inhabiting Michael Polanyi's "Republic of Science?"[3]

1. For a discussion of the uses of tacit knowledge, see Collins 2001a.
2. A start has been made on a more substantial study of "referred expertise" in Collins and Sanders 2008 (forthcoming).
3. See Thorpe 2002 for a discussion of Oppenheimer and leadership. See Collins 2004a for a detailed case study of the strains within the Laser Interferometer Gravitational Wave Observatory project as it grew from "small" to "big." Polanyi, e.g. 1962, put forward the idea of the Republic of Science.

The puzzle is also found well outside the sphere of science and technology. What should be the relationship of anthropologists to native peoples and of colonialists to the people of the countries they colonize? How well do you need to understand the people to know what is best for them? Are outsiders bound always to work from their own self-interests or frames of reference in the absence of deep understanding of the other's world? The modern academic version of this debate is found in extreme standpoint epistemologies and the politicization of academic studies of minorities. Is it the case that only blacks can do deep academic research on blacks, women on women, and the profoundly deaf on the profoundly deaf? Turning away from academia to industry, do corporate managers need to understand the work of the specialists who design and produce the products upon which their firms depend, or is it the case that all management is the same? Lord Campbell, talking of his management of London Weekend TV, once put it this way:

> You unit heads may think that managing talented producers and performers raises special problems but I have been in sugar all my life and I can assure you that the management of people in television is precisely the same as the management of sugar workers.[4]

Perhaps industries would be healthier if specialists and scientists more often found their way onto the boards of companies, perhaps not.

The "Folk Wisdom" View

The recent twist in this old debate is the "folk wisdom" view—the claim that ordinary people are wiser than experts in some technical areas. For example, consider this comment, quoted in a 1999 British report, *The Politics of GM Food:*

> "Many of the public, far from requiring a better understanding of science, are well informed about scientific advance and new technologies and highly sophisticated in their thinking on the issues. Many 'ordinary' people demonstrate a thorough grasp of issues such as uncertainty: if any-

---

4. Quoted in Muir 1997, 324–25. Muir is explaining the mass defection of program makers from London Weekend Television when, in the 1970s, the Board of Directors sacked their talented boss.

thing, the public are ahead of many scientists and policy advisors in their instinctive feeling for a need to act in a precautionary way."[5]

It is interesting and surprising to compare the folk wisdom view with the other instances of the debate about the value of specialist expertise listed above—there are some strange bedfellows! For example, under the folk wisdom view it is the ordinary person that is said to understand the closed and narrow world of science merely by observing its surface—just as the colonialists and Victorian anthropologists were said to be able to understand the world of the natives without direct experience. Here the ordinary people are thrust into a position like that of the elite, Oxbridge-trained, amateurs of the pre-Fulton Report British Civil Service—"we do not need experts among us, good thinking is sufficient."[6] Likewise, we find the ordinary person being given the same role in respect of expertise as that of Senator Proxmire and Lord Campbell. Finally, we find, implicitly, that the ordinary people are not in need of the specialist experience championed by those who believe in extreme standpoint epistemologies when it comes to understanding and researching ethnic or other minority groups. Could it be that under this implicit model it is the ordinary person represented by Alf Garnett ('*Til Death Do Us Part*) or Archie Bunker (*All in the Family*), who must be taken to hold a robust, common-sense view of minorities, in no need of refinement from arcane academic analysis?

Strangely, all this runs counter to another central theme in the social analysis of science—the idea that genuine understanding involves tacit knowledge. Tacit knowledge is the deep understanding one can only gain through social immersion in groups who possess it. Indeed, it has been claimed that most of those distant from the research front of science live their lives in a world of false certainties—sometimes positive, sometimes negative. "Distance lends enchantment" is the phrase that has been applied to this perspective, because from far away it is hard if not impossible to discern the complexities which lead scientists to be cautious in making claims. Just as one sees only the figure and not the blobs and smears of paint that make it up as one steps back from a painting, the dis-

5. *The Politics of GM Food: Risk, Science and Public Trust* (Swindon: Economic and Social Research Council, ESRC Special Briefing No. 5, October 1999), quoted at page 4. We will examine this report in greater detail in the Conclusion. For a description of more sources of the folk wisdom view in modern social studies of science, see Kusch 2007.

6. See Fulton 1968 and Hennessy 1989.

tanced view of science presents an illusory sharpness. It is the idea of tacit knowledge developed through practice that lies at the core of our analysis of expertise and upon which the Periodic Table of Expertises rests. There is an evident conflict between the idea of tacit knowledge and the folk wisdom view.

Another guiding principle of this book is that humans are unique in the way they share tacit knowledge. Humans have an ability to develop and maintain complex bodies of tacit knowledge in social groups that is not possessed by non-human entities.[7] Humans can share this knowledge with new members of the group in ways they cannot explicate. The point can, once more, be made with the example of natural language-speaking. No non-human entity has the fluency in natural human languages that comes naturally to most humans. Humans learn a natural language by being immersed in the social group to which the language belongs, and they maintain their fluency by continual social interaction with the group. This way of acquiring and maintaining fluency in a human language is beyond the capacity of any known nonhuman entity, whether artificial intelligence or biological organism. Mere physical proximity to the relevant groups is not enough. Dogs, cats, chimpanzees (arguably), humans with brain damage, and those who were not taught language early enough in their lives, and computers, are regularly exposed to language-speaking groups, yet they do not become fluent. Most of the levels of expertise set out in the Periodic Table are analogous to natural languages; they are learned through social interaction and they are maintained through social interaction. As with language, so with the expertises analogous to language—coming to "know what you are talking about" implies *successful* embedding within the social group that embodies the expertise.

The theoretical ideas we are putting forward are easy to caricature, so here are some caveats:

1. We are trying to analyze expertise and define types of expert. This is not the same as actually making technical decisions or even helping the class of technical experts in their decision-making except insofar as defining types of expertise and types of expert may contribute to a judgment. Our own technical expertise is a "meta-expertise." It is about making judgments about experts and expertise from within our own expertise—a philosophically orientated social science.

7. The popular "Actor/Actant Network Theory" ignores all this.

2. We are not claiming that all technical decisions in the public domain should simply be left to experts or even that we always know who the experts are. All such decisions take place in a political context that bears on both problems. We are saying that we must leave a logical space for certain types of expertise to be recognized independently of politics.

3. To analyze expertise is not to establish Philosopher Kings or their expert equivalents. First, this is because, at best, we are talking only about technical expertises. Political choice is a domain which we do not consider to be technical. Failing to maintain a distinction between science and technology, on the one hand, and politics on the other, leads to the stark choice between technological populism, in which there are no experts, and fascism, in which the only political rights are those gained through supposed technical expertise. In this book we try to demarcate a domain of expertise but only within the technical component of realms which have a technical component. Democracy cannot dominate every domain—that would destroy expertise—and expertise cannot dominate every domain—that would destroy democracy.

4. We make no claim here to be able to distinguish between experts who have integrity and those who acquire expertise, genuine or assumed, in order to pursue a disguised self-interest. It goes without saying that we condemn such persons.

5. The science establishment has for too long had an unhealthy monopoly on scientific and technological judgment and on the way questions that contain a technological element have been framed. The inappropriateness of this monopoly ought to obvious even within the technical domain: the technical point is that, at best, scientific knowledge takes a long time to make and therefore scientists are often pressed to make authoritative decisions on technical matters before there is any consensual scientific knowledge on which to base them. In general, *the speed of politics exceeds the speed of scientific consensus formation.* As a result too much greed for scientific authority is bad for science, forcing scientists to act in scientifically inauthentic ways. Yet, too often, science's spokespersons have claimed to be the custodians of universal truths akin to those offered by morality or religion. Ownership of the universal and eternal has to be given up if the defense of science is to have integrity. We have to understand how to use a *fallible* science and technology, *as science and technology,* in the long and sometimes indefinite run-in to what comes to count as scientific fact. The problem is not one of epistemology. It is simply a matter of how to use science and technology before there is consensus in the technical community.

The Problem of Legitimacy and the Problem of Extension

In retrospect one can see that logic and science are always too fragile for the purposes of policy-making because they are too pure. Their strength is like that of glass—hard and rigid but vulnerable to a single dislocation—and cracks are always appearing. For the material of technical policy-making, even in those realms where we are more rather than less confident about being able to demarcate the technical from the political, we need something mixed and tangled. We need something that resists even after it has been repeatedly battered and dented. We need something like a hawthorn hedge, or reinforced concrete, or earthworks; smash into such structures and though you make take lumps out of them they still stand. Our raw material is expertise mixed with experience. Our goal is to analyze this aggregate substance.

To re-express these sentiments in terms of the immediate problems that face Western societies, we want to ease the tension between the "Problem of Legitimacy" and the "Problem of Extension." The Problem of Legitimacy is about how we can continue to introduce new technologies in the face of the widespread and growing distrust of certain areas of science and technology.[8] The proper and sensible solution has been to extend the involvement of the public in the decisions. Greater dialogue between the science "establishment" and the public is now routinely demanded along with increased participation in science and technology decision-making. For example, in Britain, the Report of the House of Lords Science and Technology Committee recommended:

> That direct dialogue with the public should move from being an optional add-on to science-based policy-making and to the activities of research organisations and learned institutions, and should become a normal and integral part of the process.[9]

8. The problem of legitimacy has to be understood against a background of survey data that suggests that public perceptions of science are generally positive. It is wrong to think of "the public" as an undifferentiated mass with a monolithic view. The sense of crisis applies only to certain salient cases where the public is deeply and immediately implicated. Lawless (1977) discusses 45 cases of controversial science in the USA that occurred between 1948 and 1973 and lists many more, and the problem seems to be becoming more salient.

9. House of Lords 2000, paragraph 5.48. See also the *Science and Society* report (House of Lords Science and Technology Committee 2000), and the White Paper on European Governance (European Commission 2001), which has prompted work under the European Unions Framework 6 "Science and Society" program. For example, the "Science and Governance" theme has produced guidelines for use of expert advisers (European

Greater involvement of the public has, however, given rise to the "Problem of Extension": how do we know how, when, and why, to limit participation in technological decision-making so that the boundary between the knowledge of the expert and that of the layperson does not disappear? Perhaps an analysis of expertise will help to resolve the Problem of Extension while not destroying the more well-worked-out attempts to solve the Problem of Legitimacy.

Scientism

Our argument is founded in the critique of science and technology that has been developed, most notably in the social sciences, over the last three decades. Nevertheless, we defend science against those who do not share our view that it still has a special epistemological warrant and a central and vital place in Western democracies. The social psychology and politics of polarization has and will ensure that we will be read as putting forward a view that is not ours. We have and will be said to be putting forward a pro-science view redolent of the 1950s. Thus some of the arguments put forward here have been referred to as "scientism," and this gives us an opportunity to try once more to clarify what we are doing. For our purposes we can define four kinds of scientism:[10]

- *Scientism1:* An overpedantic cleaving to some canonical model of scientific method or reasoning.
- *Scientism2:* Scientific fundamentalism: a zealot-like view that the only sound answer to any question is to be found in science or scientific method.
- *Scientism3:* The view that narrowly framed "propositional questions" posed by scientific experts are the only legitimate way to approach a debate con-

---

Commission 2002) and has plans to promote the involvement of citizens in "dialogue and participation." Further information about the program is available at: http://europa. eu.int/comm/research/science-society/index_en.html (accessed 17 December 2003). Guidance on how government departments should put these principles into practice is given in the Office of Science and Technology publication *Guidelines 2000* (OST 2000) and in the *Code of Conduct for Written Consultations* originally produced by the Cabinet Office in 2000 and updated in 2005 (Cabinet Office 2000, 2005). In the USA the Loka Institute is a nonprofit organization that campaigns for more socially responsive science and technology policies. It organized the first "consensus conference" in the USA (e.g., see Guston 1999) and has run citizen panels on "Telecommunications and the Future of Democracy" (Sclove 1997). See also http://www.loka.org (accessed 17 December 2003).

10. Wynne (2003) applies the scientism label to the kind of work presented here. There are, of course, many other ways of classifying kinds of scientism. The term has a strong overlap with some uses of positivism, for which Halfpenny (1982) finds twenty-one variants.

cerned with science and technology in the public domain; this goes along
with blindness to the political embeddedness of such questions.
- *Scientism4:* The view that science should be treated not just as a resource,
  but as a central element of our culture.

Insofar as the position put forward in this book is scientistic, it is a matter of scientism4. The other three kinds of scientism are rejected. What
is included in scientism4 is a preference for the norms and culture of
evidence-based scientific argument. This preference cannot be relin-
quished in the process of policy-making without giving up much more
than most of the readers of this book would want to surrender. (In chapter 5 we will also define "artism.")

The Structure of the Book

In chapter 1 we begin the analysis of expertise. It starts with the Periodic Table of Expertises. The table is a classification of the expertises that
individuals might draw on when they make technical judgments. At the
outset of chapter 1 there is a short overview of the structure of the table.
Chapter 2 continues the exploration of the Periodic Table, concentrating
on meta-expertises—expertises used to judge other expertises. At the end
of chapter 2 the distinctions developed in the table are summarized.

The next two chapters of the book are a detailed exploration of one
new category of expertise, "interactional expertise." Chapter 3 is a philosophical examination of the idea of interactional expertise, contrasting
our position with that of some other approaches, in particular the notion
of "embodiment" as developed by Hubert Dreyfus. The notion of interactional expertise gives an importance to language communities that is
overlooked by those who stress the importance of the individual body in
the development of expertises.

Chapter 4 reports some "imitation game" experiments designed to
explore the notion of interactional expertise. We compare the linguistic abilities of the color-blind and the "pitch-blind"—those who do not
have perfect pitch. These experiments explore the "strong interactional
hypothesis"—that those with maximal interactional expertise are indistinguishable from those with contributory expertise in linguistic tests.
Chapter 4 finishes with an indication of the meaning of this finding in
the case of gravitational wave science.

Chapters 3 and 4 indicate the range of discussion and investigation
opened up by a realist theory of expertise. Though this book concentrates

on interactional expertise, in due course the other categories could be explored in similar depth.

The final chapter, chapter 5, returns to the more speculative themes opened up in this introduction. Having developed a realist theory of expertise we ask how it could be used in technological decision-making in the public domain. It could not be used without separating the sciences and technologies from expertise-laden activities which are not sciences and technologies. Therefore the last chapter concentrates on new demarcation criteria between science and the arts, science and politics, and science and pseudo-science. We hope this chapter will revitalize the debate about the difference between science and technology and other cultural endeavors.

CHAPTER ONE

The Periodic Table of Expertises 1: Ubiquitous and Specialist Expertises

The Periodic Table Introduced

Table 1 is the Periodic Table of Expertises—a table of the expertises that might be used when individuals make judgments. The table is drawn in two dimensions, but every now and again, as indicated below, a third dimension would be useful. This chapter deals mainly with the first three rows. The next chapter deals mainly with the bottom two rows— Meta-Expertises and Meta-Criteria. We begin, however, with a summary explanation of the whole table that can act as a map or "ready reference"—a quick reminder of the whole structure and the meaning of any particular category. At the end of chapter 2 we will provide another summary for those who have read through the details.

Working from the top of the table, *ubiquitous expertises* are those, such as natural language-speaking, which every member of a society must possess in order to live in it; when one has a ubiquitous expertise one has, by definition, a huge body of tacit knowledge—things you just know how to do without being able to explain the rules for how you do them. This row of the table also includes all those expertises one needs to make political judgments. Below this line the table is exclusively concerned with technical expertises—those that have science and technology content.

Dispositions are not very important to the conceptual structure of the table; they are personal qualities—the ones we discuss are linguistic fluency and analytic flair.

The next row deals with the *specialist expertises*. Low levels of specialist expertise are better described as levels of knowledge—like knowledge of the kind of facts needed to succeed in general knowledge quizzes. One may be able to recite a lot of such fact-like things without being able to do

Table 1: The periodic table of expertises

UBIQUITOUS EXPERTISES					
DISPOSITIONS				Interactive ability / Reflective ability	
SPECIALIST EXPERTISES	*UBIQUITOUS TACIT KNOWLEDGE*			*SPECIALIST TACIT KNOWLEDGE*	
	Beer-mat knowledge	Popular understanding	Primary source knowledge	Interactional expertise	Contributory expertise
				Polimorphic / *Mimeomorphic*	
META-EXPERTISES	*EXTERNAL* (Transmuted expertises)		*INTERNAL* (Non-transmuted expertises)		
	Ubiquitous discrimination	Local discrimination	Technical connoisseurship	Downward discrimination	Referred expertise
META-CRITERIA	Credentials		Experience		Track record

anything much as a result except succeed in quizzes. Three low levels of expertise are listed on the left-hand side of the specialist expertise rows of the table. It is important to note that acquiring low levels of expertise seems like a trivial accomplishment only to those who already possess ubiquitous expertise; acquisition of even these low levels rests on the prior acquisition of a vast, but generally unnoticed, foundation of ubiquitous expertise.

To acquire higher levels of specialist expertise, more than ubiquitous expertise is needed. To go further along row three it is necessary to immerse oneself in a domain so as to acquire *specialist tacit knowledge*, not just learn more facts or fact-like relationships. Two categories of higher level expertise are found at the right hand end of the specialist expertise rows. The highest level is *contributory expertise*, which is what you need to do an activity with competence. Just below this, however, is *interactional expertise*, which is the ability to master the language of a specialist domain in the absence of practical competence. The idea of interactional expertise is immanent in many roles, from peer reviewer to high-level journalist, not to mention sociologist or anthropologist, but it seems not to have been discussed before in an explicit way. A good proportion of the book is taken up with explaining the notion of interactional expertise because it is a new concept.

Moving down to the fourth row we encounter *meta-expertises*. The first set of two meta-expertises are the prerogative of judges who, not possessing the expertise in question, make judgments about experts who do possess it. This is done by judging the experts' demeanor, the internal consistency of their remarks, the appropriateness of their social locations, and so forth. These are "transmuted expertises" because they use social discrimination to produce technical discrimination. The first kind of discrimination depends on the kind of ubiquitous expertise one gains in a democratic society as one learns to choose between politicians, salespersons, service providers, and so forth. The second kind of discrimination depends on local knowledge about those around you. The second set of three meta-expertises do not depend on transmutation, as they are based on possessing one level or another of the expertise being judged. *Technical connoisseurship* is like the expertise of art critics or wine buffs who, crucially, are not themselves artists or wine-makers. The middle of the three categories relates to what we most naturally think of as skillful judgment—where one specialist judges another. There are three directions in which this middle category of judgment can be made: an expert can judge someone who is still more expert, an expert can judge someone equally expert, or an expert can judge someone less expert. Mostly experts think they are pretty good at judging in any of the three directions, but we argue that only the downward direction is reliable, the other directions tending to lead to wrong impressions of reliability or irresolvable disputes. The one reliable category which appears in the table is, therefore, labeled *downward discrimination*. *Referred expertise* is the use of an expertise learned in one domain within another domain. In the chapter we use examples drawn from the management of large scientific projects, where a manager moves from one to another, to illustrate the concept.

The final row of the table refers to the criteria that outsiders try to use to judge between experts to avoid having to make the more difficult kind of judgments described above. They can check expert's qualifications, they can check expert's track records of success, or, what we argue is the best method of the criterion-based judgment, they can assess the expert's experience.

Ubiquitous Expertises

We now move on to a more detailed analysis of each category of expertise starting with the top row. Ordinary people are talented and skillful almost beyond comprehension. We can say "almost beyond comprehension" with confidence because a lot of very clever people have tried to encapsulate the talents of ordinary people in computer programs, entirely failing to realize how hard a task it would be.[1] What we will call "ubiquitous expertises" include all the endlessly indescribable skills it takes to live in a human society; these were once thought of as trivial accomplishments.[2] For any specific society, its "form of life" or "culture" provides, and is enabled by, the content of the ubiquitous expertises of its members. Fluency in the natural language of the society is just one example of a ubiquitous expertise. Others include moral sensibility and political discrimination. These are abilities that people acquire as they learn to navigate their way through life. In the case of ubiquitous expertises the Problem of Extension ceases to have any practical significance because almost everyone is a genuine contributory expert.[3] Thus, when we say that the folk wisdom view is often misplaced we do not mean that ordinary people do not have expertise, we mean only that the ubiquitous expertise of ordinary people should not be confused with the expertise of technical specialists. What

1. We have already mentioned how this mistake was made by the pioneers of natural-language-handling computers.

2. For a discussion of the fact that the so-called unskilled persons who work in McDonald's restaurants actually have a huge depth of ubiquitous expertise, see Collins and Kusch 1998, chap. 8. Here is a clear case where the lack of scarcity value of an expertise has been taken to mean that it is easy.

We refer to *ubiquitous* expertises because the antonym of esoteric is "exoteric," which the Chamber's Dictionary defines as "intelligible to the uninitiated; popular or commonplace." This definition renders the word "exoteric" inappropriate when paired with expertise because by definition you cannot have expertise which is intelligible to the uninitiated: expertise is available only to the initiated or experienced. Our distrust of the term "lay expertise" also has its roots in the definition. The Chambers Dictionary defines a layman as follows: one of the laity; a nonprofessional person; someone who is not an expert. But all laypersons possess ubiquitous expertises.

The Aristotelian term *phronesis*, being some sort of combination of prudence and wisdom—a practical wisdom in a moral setting—captures part of the notion of ubiquitous expertise but not that part represented by language-speaking and the like.

3. The success of lawyers suing firms such as MacDonald's for selling over-hot coffee, and the consequent growth of warnings and safeguards surrounding every consumer good, is patronizing—treating the public as incapable of learning the rules of ordinary living through the normal processes of socialization.

Technical expertises can be ubiquitous in certain societies and not in others—for example, the ability to wire a plug or mend a fuse. In the Society of Social Studies of Science meeting in Pasadena in 2005, Wiebe Bijker suggested that the Dutch population as a whole had sufficient expertise in dam-building to contribute to public debates about new projects in contrast to what everyone agreed was a low level of expertise in the working of levees among the general population of Louisiana. Whether this claim is true or not remains to be proved, but it is an intriguing suggestion that is central to the concerns of this book.

we are arguing is that we must preserve a logical space for expertises that are not the property of the general public; it is impossible for the general public to have expertise in every specialist technical domain even though they have a vast store of ubiquitous expertise.

Before passing to specialist domains it is worth noting that just because *some of the things* we can all do are hugely skillful it does not mean that *all the things* we can all do are hugely skillful and this includes things of which we have great experience. For example, one might have huge experience of lying in bed in the morning, but this does not make one an expert at it (except in an amusing ironic sense). Why not? Because anyone could master it immediately without practice, so nothing in the way of skill has been gained through the experience.

Now we turn to the central question of how much expertise in specialist domains it is possible for ordinary people to have. To answer the question we need to think about ways of having specialist expertise. We can construct a rough ladder of knowledge about, or expertise in, specialist domains. No doubt the bottom of the ladder could be divided up in more or less ways and along different dimensions, but we need to start with something.

As explained, our model is human-centered. At some stage all human expertise touches on tacit knowledge, that is, an understanding of rules that cannot be expressed. It is the inexpressibility of the rules of ubiquitous expertises that make them so hard to capture in computers. The idea of tacit knowledge will be discussed further in chapter 3, but for now we need to note only that tacit knowledge enters into knowledge acquisition in two ways. Some kinds of knowledge acquisition amount to the acquisition of additional specialist tacit knowledge; other kinds of knowledge acquisition involve the exercise of tacit knowledge *in the course of* the acquisition of information. In the second kind of knowledge acquisition, the tacit knowledge used is that found in ubiquitous expertises. For example, the exercise of the ubiquitous expertise associated with language can be used to acquire new information (explicit knowledge) through reading or listening without interactive discourse. The first rungs of a ladder of specialist expertise involve only ubiquitous tacit knowledge used in this way. The higher level rungs require immersion in the tacit knowledge of the specialist domain so that more tacit knowledge can be acquired. For example, mastery of even a widely distributed tacit-knowledge-laden expertise such as car-driving needs practice at car-driving and internalization of the unspoken rules of road-craft. Likewise, becoming a full-blown specialist in a scientific or technical domain requires immersion in the society of the domain specialists. This gives us an initial division of types of expertise: those that can be acquired using only ubiquitous tacit knowledge and those that involve specialist tacit knowledge.[4]

Expertises Involving Only Ubiquitous Tacit Knowledge

Beer-Mat Knowledge

So far we have established than an essential component of any ladder of human expertise is *ubiquitous expertise.* The first rung of the specialist ladder is what we will call "beer-mat knowledge." Consider the following explanation of how a hologram works:

> A hologram is like a 3 dimensional photograph—one you can look right into. In an ordinary snapshot, the picture you see is of an object viewed from one position by a camera in normal light.
>
> The difference with a hologram is that the object has been photographed in laser light, split to go all *around* the object. The result—a truly 3 dimensional picture!

This explanation, found on a beer mat made for the Babycham company in 1985, appears to give an answer to the question "What Is a Hologram?" It is capable, presumably, of making at least some people feel that they now know more about holograms. The words on the beer mat are not simply nonsense nor could they be taken to be, say, a riddle or a joke. Presumably there are people now alive who have studied the beer mat and, if asked: "Do you know how a hologram works?" would reply: "Yes," whereas immediately before they had read the beer mat they would have answered: "No," to the same question. So what increment in expertise does someone have in consequence of perusing the beer mat?

Let us investigate by analyzing another such thing that one might know, the rule for the move of the bishop in chess. The rule, and we might well read it on a beer mat or something similar, is "the bishop may move, only diagonally, any distance, backwards or forwards." But it is possible to know this rule in more than one way. One might know

---

4. Notice that an expertise such as skillful car-driving is very widely distributed but it is not ubiquitous. Car-driving is not learned integrally with learning to live in society but needs specialist training and the specialist tacit knowledge that goes along with it, even though nearly everyone in certain societies can do it.

it in the same way as an observant Jew or religious Catholic might know how to recite certain prayers in Hebrew or Latin respectively but without knowing their meaning. Thus, knowing how to "chant" the bishop's move might enable one to score a point in a board game such as Trivial Pursuit, which is intended to discriminate between levels of general knowledge. Crucially, knowing the bishop's move in that way does not imply that one knows much about what it might mean. For example, you can know it in the beer mat/Trivial Pursuit way without knowing that the term "any distance" within the rule is to be measured in squares on the chess board and that it can never be more than seven squares nor that "any distance" means only so long as the path is not blocked by another piece, nor that the restriction to the diagonal implies, on a chessboard, that the bishop is restricted to squares of only one color (nor even that there are two colors on the chessboard). In short, knowing the bishop's move in a beer-mat-knowledge kind of way does not enable one to *do* anything much that one would not be able to do if one did not know it (other than scoring points in general knowledge tests). Knowing the rule for the bishop's move in the context of chessboards and the game of chess is a rather different thing even for a novice chess-player than for a Trivial Pursuit player. The novice player who knows the rule knows how to move the bishop on a chessboard. (However, it is only with further experience that the novice learns how to recognize when, and in what circumstances, moving the bishop might be a good idea. And it takes the skill of a chess master to see that making *this* move with the bishop is the turning point of the game.)

Going back to the hologram, the explanation on the beer mat does *not* enable the naive reader to do anything such as make a hologram, or debate the nature of holograms, or to correct anyone's mistakes about the nature of holograms, or to make a sensible decision about the long-term dangers associated with the unrestrained spread of holograms, or convey any information about holograms other than the formula itself.

Popular Understanding of Science

Moving to the next rung of the ladder, much superior to beer-mat knowledge is what we will call "popular understanding."[5] Popular understanding can be gained by gathering information about a scientific field from the mass media and popular books. It is the kind of understanding to

5. We are grateful to Matthew Harvey for helping develop this popular understanding category.

which bodies such as the Royal Society's Committee for the Public Understanding of Science (COPUS) once directed its efforts.

Popular understanding does involve a deeper understanding of the *meaning* of the information than does beer-mat knowledge. For example, it may be possible to make some inferences from popular understanding of science of the kind "antibiotics will not cure viral diseases, influenza is a viral disease, antibiotics won't cure influenza," or "the element is completely enclosed in my electric kettle whereas heat is wasted when I boil water on the gas stove, so the electric kettle uses less energy to boil the same amount of water than the gas stove provided not too much energy is wasted converting gas to electricity in the power station." Popular understanding of science is also transmissible from one person to another to a certain extent—transmissible as a set of ideas rather than a set of formulae.

In the case of a long-settled science the difference between a deeper understanding of science and technology and a popular understanding is not very important in terms of public decision-making; where the science is settled, the difference between scientific knowledge as revelation and deep scientific understanding has little impact on the conclusions reached because both give rise to the same judgments. Where the science is the subject of a dispute, however, the difference is of the essence. The last three decades of social studies of science have shown us that, in disputed science, detail, tacit knowledge, and unspoken understandings of who is to be trusted among those who work in the esoteric core of the science are vital components of decision-making at the technical level. Popular understanding hides detail, has no access to the tacit, and washes over scientists' doubts. The consequence, well-established in the sociology of scientific knowledge, is summed up in the phrase mentioned in the introduction: "distance lends enchantment." Special cases aside, the more distant one is from the locus of the creation of knowledge in social space and time the more certain will the knowledge appear to be. This is because to create certainty, the skill and fallible effort that goes into making an experiment work, or a theory acceptable, has to be hidden; if the human activity that is experimentation is seen clearly, then it is also possible to see all the things that could be wrong.[6] Any rede-

---

6. For "distance lends enchantment," see Collins 1992. For a modification, see MacKenzie 1998. Ludwik Fleck, who was a sociologist of scientific knowledge before the term was invented, wrote in the 1930s:

Characteristic of the popular presentation is the omission both of detail and especially of controversial opinions; this produces an artificial simplification [and] . . . the apodictic

scription of events in the core of science, even when it is designed for a professional audience, is bound to simplify; when the description is for a popular audience, it will simplify more brutally. But sound judgments, or at least *informed* judgments, in disputed science must take account of many more of these uncertainties than popular understanding allows for. For this reason, in the case of disputed science, a level of understanding equivalent to popular understanding is likely to yield poor technical judgments.

The problem of judgments based on popular understanding applies whether the conclusion is positive or negative—whether the consumers of the simplified version accepts everything they read and hear (for example, they might accept that Stephen Hawking's utterances about black holes are revealed truth), or rejects the claims (for example, they might be certain that everything the government says about the safety of vaccines is false). Both kinds of interpretation of evidence are strengthened and reinforced by distance and by the "narrow bandwidth" of the media which provide popular understanding.[7]

One of the troubles with the old officially sponsored approach to popular understanding is that it does not distinguish between consensual science and disputed science. It tends to present even disputed science as revealed knowledge emerging from a unified community of experts. This converts any genuine effort at increasing public *understanding* into propaganda.[8] The obvious danger, even for those keen on propaganda, is that for each positive piece of propaganda there is a negative one which will be grasped with equally unmodulated certainty.

In sum, popular understanding is a big step up from beer-mat knowledge but a long way from deep understanding of scientific matters. The

---

valuation simply to accept or reject a certain point of view. Simplified, lucid, and apodictic science—these are the most important characteristics of exoteric knowledge. *In place of the specific constraint of thought by any proof, which can be found only with great effort, a vivid picture is created through simplification and valuation.* (Fleck 1979 [1935], 112–13, emphasis in original)

7. Treatments that turn on the establishment of scientific knowledge as a matter of literary transformation as work passes from laboratory to the wider world (for example, Latour and Woolgar 1979) accurately describe the way certainty increases as the bandwidth narrows (for example, details of the time, place, and personal involved in an experiment are successively removed as accounts enter more public domains). What they do not explain, however, is how the certainty comes to be that the account is unproblematically true, on the one hand, or unproblematically false, on the other.

8. A high priest of this approach is Lewis Wolpert, once chair of COPUS, whose book *The Unnatural Nature of Science*, stressed just how different a scientific grasp of matters was to a commonsense appreciation.

gap between popular understanding and deep understanding is not so important where the science is settled and consensual, but it is very important where science is disputed. Not by chance, wherever there is a serious public debate involving science, the science is nearly always disputed, so the enchantment brought about by distance from the research front, whether negative or positive, is crucial.

Primary Source Knowledge

The next step after popular understanding is the kind of knowledge that comes with reading primary or quasi-primary literature. We will call it "primary source knowledge." Nowadays the Internet is a powerful resource for this kind of material. But even the primary sources provide only a shallow or misleading appreciation of science in deeply disputed areas, though this is far from obvious: reading the primary literature is so hard, and the material can be so technical, that it gives the impression that real technical mastery is being achieved. It may be that the feeling of confidence that comes with a mastery of the primary literature is a factor feeding into the folk wisdom view.[9]

Actually, it can be shown that what is found in the literature, if read by someone with no contact with the core-groups of scientists who actually carry out the research in disputed areas, can give a false impression of the content of the science as well as the level of certainty. Many of the papers in the professional literature are never read, so if one wants to gain something even approximating to a rough version of agreed scientific knowledge from published sources one has first to know what to read and what not to read; this requires social contact with the expert community. Reading the professional literature is a long way from understanding a scientific dispute.[10] The question, then, even for those who read the journals in which primary research findings are published, is whether their knowledge matches the Trivial Pursuit player's, the chess novice's, the experienced chess player's, or the chess master's understanding of the

9. A familiar image is today's informed patient visiting their doctor armed with a swathe of material printed from the Internet. While this kind of information gathering, especially in the context of a support or discussion group, can be valuable, it is important not to lose sight of what sociologists have shown: a great deal of training and experience is needed to evaluate such information. For a discussion of expertise in the medical context, see Collins and Pinch 2005.

10. Thus, there are published physics papers, making potentially momentous claims, that are known by the initiated to be of no scientific importance (see, for example, Collins 1999, 2004a).

bishop's move. Our claim is that in the case of scientific disputes primary source knowledge is not much better in respect of the science than a chess novice's understanding in respect of the bishop's move.

Expertises That Involve Specialist Tacit Knowledge

Over the last half-century, the most important transformation in the way expertise has been understood is a move away from seeing knowledge and ability as quasi logical or mathematical and toward a more wisdom-based or competence-based model. As has been intimated, expertise is now seen more and more as something practical—something based in what you can do rather than what you can calculate or learn. This shift has been in part inspired by ideas coming from phenomenological philosophers such as Heidegger and Merleau-Ponty. Polanyi, who invented the term "tacit knowledge," has also been influential, especially among scientists and philosophers of science, while for sociologists of science the main influence has been Wittgenstein's idea that the meaning of a concept can be understood only through its use; it is the use of a concept that establishes its meaning, rather than any kind of logical analysis or a dictionary definition.[11] The Wittgensteinian frame of mind (as interpreted here) leads us to expect to find specialist knowledge located in specialists' practices rather than in books.[12] Mastering a tacit knowledge-laden specialism to a high level of expertise, whether it is car-driving or physics, ought, then, to be like learning a natural language—something attained by interactive immersion in the way of life of the culture rather than by extended study of dictionaries and grammars or their equivalents. The first three categories of expertise, beer-mat knowledge, public understanding, and primary source knowledge, might be said hardly to enter the category of specialist expertise at all because they do not involve much in the way of mastering the tacit knowledge belonging to

11. Wittgenstein 1953. Wittgenstein's writings are somewhat aphoristic and open to many interpretations. The interpretation adopted here is that of Winch 1958, and also coincides with that of Bloor 1973 and 1983.

12. We must not pass this point without noting that the logic of what is currently the most dominant trend in science studies, the work of Bruno Latour and Michel Callon around so-called "actor network theory," includes absolutely no role for expertise or any other special property that pertains to human societies or the particular capabilities of humans. On the contrary, actor network theory takes even inanimate objects to be ontologically indistinguishable from humans. Thus, while we talk of a change in our understanding of knowledge, this does not apply to the dominant part of science studies as it is practiced today. For a critique of actor network theory, see Collins and Yearley 1992.

the subject matter of the domains; the acquisition of the first three kinds of knowledge (though it depends on ubiquitous expertises), involves reading rather than immersion in the specialist culture. "Enculturation" is the only way to master an expertise which is deeply laden with tacit knowledge because it is only through common practice with others that the rules that cannot be written down can come to be understood.

What is new about our analysis of expertises learned through immersion in a culture is that we split them into two. The traditional category of ability to perform a skilled practice we call "contributory expertise." *Contributory expertise*, as its name suggests, enables those who have acquired it to *contribute* to the domain to which the expertise pertains: contributory experts have the ability to *do* things within the domain of expertise. This is the traditional way of thinking about this kind of expertise, and we discuss it first before moving on to the new category of *interactional expertise*, an idea which we consider to be a significant contribution to the understanding of expertises in general.

Contributory Expertise

The five-stage model of acquisition of contributory expertise is one of the more well-known and influential schemas, and we will use it to stand for all the important approaches—those that stress the importance of the "internalization" of physical skills.[13] It could be usefully represented on a third dimension of the table. According to the five-stage model, only at the early stages of skill acquisition is there need for calculation or even self-conscious rule-following (the left-hand side of the specialist expertise row in our table); self-conscious application falls away as a skill becomes "embodied"; this is essential for efficient performance. The five stages can be exemplified by the process of learning to drive a car:[14]

- Stage 1 is the *novice*. The novice driver will try to follow explicit rules and as a result the performance will be labored, jerky, and unresponsive to changes in context. The skill will be exercised "mechanically," following rules such as "change gear when the car reaches 20 mph as indicated on the speedometer." These are "context-free" rules, because in applying

13. Dreyfus and Dreyfus 1986. There are, of course, other kinds of approach to expertise and apprenticeship, many of which emerge from the field of education research. See for example, Ainley and Rainbird 1999; Coy 1989; Pye 1968; Lave 1988; and Lave and Wenger 1991.

14. Dreyfus and Dreyfus 1986, 21–36.

them the learner does not take into account the nuances implied by different conditions of application.

- Stage 2 is the *advanced beginner*. As more and more of the skill is mastered, however, more unexplicated features of the situation start to play their part in the performance, such as using the sound of the engine as an indicator of when to change gear, which will in turn mean gear changes at different speeds according to whether, say, the car is going uphill or down.
- Stage 3 is *competence*. Here the number of "recognizable context-free and situational elements" becomes overwhelming, and expertise becomes much more intuitive rather than calculating. "Problem solving" is no longer the predominant motif.
- Stage 4 is *proficiency*. The proficient driver recognizes whole problem situations "holistically" in the same way as the advanced beginner recognizes specific features of the environment. In the advanced beginner stage it is, say, the sound of the engine that is recognized from experience; in the proficient stage it would be a complete traffic scenario. Nevertheless, some elements of conscious choice and analysis remain to guide the proficient driver's decisions.
- Stage 5 is *expertise*. When expert status is achieved, complete contexts are unselfconsciously recognized and performance is related to them in a fluid way using cues that it is impossible to articulate and that if articulated would usually not correspond, or might even contradict, the rules explained to novices. Hence the common experience of driving a familiar journey to work and being unable to remember anything about it when one arrives; something else was occupying the mind during the journey and the unselfconscious self was left to cope with controlling the car just as it normally copes with walking or chewing (again, activities about which we know nothing in a self-conscious way). It is also the case that when experts attempt to revert to a more self-conscious way of tackling a task they do it less well—like the mythical centipede that would trip over its legs if it thought about where it was putting them. In sum, skills practiced by individuals have to be "internalized" if they are to be practiced efficiently.[15]

---

15. For pedestrians, one may think of the way we learn to cross the road. Explicit rules—"look left, look right, look left again and if nothing is coming walk rapidly across"— disappear as they become absorbed into the generalized skill of crossing roads; this is a skill that suddenly reappears, and has to be relearned starting with a conscious routine, when we go to a country where they drive on the other side of the road. Knowing how to cross the road is known by the novice as a set of explicit and fixed rules, but by the experienced road-crosser as an unexplicated skill which is acted out in different ways as each new and

The five-stage model would be represented in a three-dimensional table by columns coming out of the page wherever a practical skill was under examination, but it has no bearing on many of the categories. A problem with the five-stage model, even as a discussion of contributory expertise, is its individualistic nature. Bicycle riding has a venerable history as an example in the debate about the nature of skill, and we will switch for a moment from cars to bikes. Michael Polanyi introduced the bicycle example, pointing out that the physics of riding a bike is exceptionally complex and counter-commonsensical and certainly of no use to those wishing to learn to ride. But imagine that our brains and nerve impulses were speeded up a millionfold: Would things change? We can ask the same question in reverse, as it were, by slowing everything down. Suppose the loss of balance happened much less quickly (as in bicycle-riding on the Moon or on an asteroid with a still lower gravitational field). The bike might fall over so slowly that there would be time to read a book of balancing instructions and follow them in the new, much slower, real time. Bike-riding would then become more like assembling flat-pack furniture: you hold the instructions in one hand and obey them without any significant time constraints.[16] The physics of bike-riding is not, then, as forbidding as it seems. Though humans cannot master it, there seems no a priori reason why a much faster non-human machine could not master it.[17]

Crucially, what Polanyi was discussing was not "bicycle-riding" but "bicycle-balancing." *Bicycle-riding* has two components: the first is balancing upright; the second is negotiating traffic. Car-driving has the equivalent two components: the first is control of the gears, steering, etc. and the second is, once more, negotiating traffic. Negotiating traffic is a problem that is *different in kind* to balancing a bike or using the clutch in that it includes understanding social conventions of traffic management.

---

unanticipated circumstance is encountered. As ability to cross the road increases, the pedestrian seems to *know* less and less about it. Actually, the experienced road-crosser uses a non-machine-like set of procedures having to do with making eye contact with the driver, and so forth. Learning to cross the road fits the Dreyfus model quite well.

16. See Collins 2007 for a discussion of bicycling on the Moon. Or think about the human-specificity of the skill of sportsmen and women. Batting at baseball or cricket would be an entirely different proposition if the batter's brain worked 1,000 times as fast. It would be a matter of hitting what would be, essentially, a stationary ball—the skills required would be more like those of stationary ball games, such as pool or golf—something far more calculative. (Collins and Kusch 1998 point out that golf-ball-striking can be done better by machine.)

17. As a matter of fact, bike-riding has been accomplished by mechanical means, but this seems to be a matter of analogue feedback from gyroscopic sensors. There seems no reason why the analogue device could not be reproduced by a sufficiently complex digital version.

These are the property of social groups; they vary from place-to-place and time-to-time. To master them requires not embodiment of the skill but being socialized into the relevant group practices.

The difference between bicycle—balancing and negotiating traffic has been described in terms of the difference between mimeomorphic and polimorphic actions.[18] Mimeomorphic actions, however complex, and however hard to master, do not turn on social understanding and can, in principle, be reproduced by mimicking fixed behaviors—though sometimes these will be too complex in practice to be accomplished. It is for this latter reason that automation of factories and so forth has to start with standardization of the whole manufacturing process, not just the replacement of individual machines within the chain of production. Furthermore, humans (working in a normal time-frame), cannot master complex mimeomorphic actions in a machine-like way for the reasons explained by Polanyi. In most cases humans have to internalize the abilities and the process of learning the new abilities in ways that appear similar to the learning of a social skill. Closer analysis, as in the case of bicycle-balancing, shows that this is a matter of the limitations of humans rather than the intrinsic nature of the expertise. It follows that sometimes machines that do not have human limitations can master the skills; we can easily imagine a very fast computer being constructed that would use the explicit physics of bike-riding along with an array of feedback devices to balance a moving bike. On the other hand, polimorphic actions, which do depend on social understanding, require that behavior fits changing social circumstances, and they cannot be mastered by machines failing a way of making machines that fit as smoothly into social life as humans.

We indicate the difference with the two boxes beneath the contributory expertise box in the Periodic Table but one might think of this distinction too as something that should be represented by a third dimension coming out of the page since a similar analysis could be conducted for every box. We will not say much more about mimeomorphic and polimorphic actions here as they are the subject of an entire book—that referred to in the last footnote. The distinction is crucial for understand the relationship between humans and machines but also for the proper understanding the relationship of the human body and brain in the acquisition of expertises.[19]

18. Collins and Kusch 1998.

19. The famous example of the breadmaking machine discussed by Nonaka and Takeuchi (1995) would also have benefited from breaking down the actions of human breadmaking into its mimeomorphic and polimorphic components. If bread, like music, is sometimes

Interactional Expertise

The overlooked second type of deeply tacit-knowledge-laden expertise is interactional expertise. This is expertise in the *language* of a specialism in the absence of expertise in its *practice*. This may seem contrary given all that we have just said about the importance of practice—of doing things—but we must look more deeply.

Why Has Interactional Expertise Been Overlooked?

To simplify, within the existing academic literature analysts tend to think of knowledge as of two kinds: the formal or propositional, on the one hand, and the informal or tacit, on the other. The formal can be expressed in rules, formulae, and facts, and can be encapsulated in computer programs, books, and the like. The informal or tacit, insofar as it is also rule-like, comes in the form of rules that cannot be explicated and are known only through their expression in action. They can be recognized as rule-like because it is easy to see when they have been broken. That is, it is easy for those who have internalized the rules, by being enculturated into the form of life that expresses them, to see when they have been broken.[20] The perennial question that emerged with particular clarity in the debate about "artificial intelligence" is whether the informal can be reproduced by sets of formal rules if the set of rules is made large enough. This question has tended to polarize analysts.

To put this another way: language, whether natural language or the language pertaining to a specialist domain, has been treated in one of two exclusive ways:

· *Informal view:* Full immersion in an entire form of life would be needed to master a language.

---

made to fit a context, then making bread using a machine is like listening to a recording of a concert rather than listening to a live performance. The former is always exactly the same, the latter varies subtly each time. In fact, even when we move to the apparently mimeomorphic aspects of breadmaking we are likely to find that the inputs and outputs will have to be more standardized than would be necessary in the case of a human breadmaker (Ribeiro and Collins 2007). For an extended discussion of the relationship between Dreyfus's analysis of expertise and that discussed here see Selinger and Collins 2007.

20. For example, I may not be able to say what the rules for proximity to others are in various societies, but with a little habituation I will be able to accomplish them and I will also soon be made to know if I break them, and can easily recognize if someone in my own society breaks them (say, by standing too close to me).

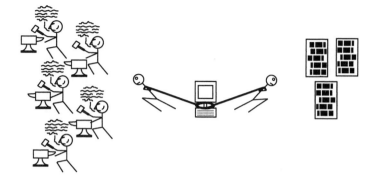

Figure 1. Is the "intelligent computer" text or practice?

- *Formal view:* Mastering the language pertaining to a domain comprises no more than the acquisition of propositional knowledge—a set of formal rules and facts gained through reading and instruction.

The second of these possibilities—that of the "formalists"—has, we believe, been thoroughly exploded. It has been shown to be wrong by theoretical analysis, and it has been shown to be wrong by the failure of the long-running experiment with intelligent computers. This experiment shows that the language of a domain, like any spoken language, consists of more than propositional knowledge. The problem this has left is that any attempt to claim that language can be mastered outside of a full-blown practical immersion in a form of life is thought, by most informalists, to amount to the claim that the formal view is true since the only possibilities on the table, as it were, are the informal and formal views. This excludes analysis of any kind of immersion within a domain that is short of full-blown practical immersion. The awkwardness of the notion of *interactional expertise* comes from the fact that it stands between the two views.

Figure 1 illustrates the point in cartoon form. On the left we have a group engaged in the practical activity and the discourse of a form of life. On the right we have a description of their activities in propositional form—in books, journal articles, and the like. In the middle we have a computer, the subject of an intellectual tug-of-war between the "formalists," who believe it can be elaborated to the point where it will fit indistinguishably into the left-hand group, and the "informalists," who believe it will never progress beyond what is, essentially, a set of propositions or symbols whose natural home is with the printed artifacts and the like on the right. The existence of this tug-of-war means that the efforts and the gaze of those who think about these matters are directed exclusively one way or the other; no one is paying attention to the space between the skilled group and the books. For example, one of the standard motifs in critiques of artificial intelligence is the mistakes made by coaches of any practical activity, the point being that the coach, whether mechanical or human, cannot express in words what the athlete or other kind of learner can only master by action.[21] The idea of interactional expertise gives us a better idea of what some human coaches might be doing and how they succeed despite the gap between language and practice. As we now see, the human coach *can* teach some things through the medium of spoken language because the coach shares some of the nonexplicit skills of the student: the shared linguistic skills can transfer mutually understood tacit meanings that would not be available to those with levels of expertise below interactional.[22]

Interactional expertise, then, is found in this middle ground between practical activity and books, computers, and so forth. Interactional expertise is, however, nearer to the informal than to the formal view. Interactional expertise is far from a set of propositions. Interactional expertise is mastery of the language of a domain, and mastery of any language, naturally occurring or specialist, requires enculturation within a linguistic community. Interactional expertise cannot be expressed in propositional terms. The computer, no current or foreseeable model of which can be immersed within a language community in a way that will allow for it to become enculturated, will have to be dragged to the right. On the other hand, the idea of interactional expertise still does not amount to the informal view—full-blown immersion in a form life. The idea of interactional expertise implies that complete fluency in the language of a specialist domain can be acquired in the *absence* of full-

21. For typical critiques of coaching see Dreyfus 1972, Dreyfus and Dreyfus 1986, and Collins 1990.

22. Coaches can also transfer tangential rules, such as "hum the Blue Danube when you swing your golf club," and "second order measures of skill." Second order measures of skill are such as: "If you are a surgeon, intending to spay a ferret, and you cannot find its uterus first time, look again but don't look more than about six times." These specific sets of rules do not comprise the skill but do give valuable guidance to a human trying to master a skill. In the same way it is good to know that (1) a human can ride a bicycle; (2) it is likely to take several hours to learn; (3) a human can learn to play the piano; (4) it is likely to take at least a year to learn; (5) a human cannot learn to fly unaided. Collins 1990, chap. 6, discusses tangential rules. Pinch, Collins, and Carbone 1996 deals with second order measures.

Figure 2. The sociologist delighted with interactional expertise

blown physical immersion in the form of life. To try to be as sharp as possible, we make a "bold conjecture": the testable "strong interactional hypothesis." The strong interactional hypothesis states that, in principle, the level of fluency in the language of a domain that can be attained by someone who is only an interactional expert is indistinguishable from that which can be attained by a full-blown contributory expert. Figure 2 illustrates the point.

Figure 2 shows the computer in its proper place, while interactional expertise occupies the left center of the middle ground. That the experts are engaged in interactional expertise is indicated by taking away their anvils, so their hammers merely identify them as contributory experts engaged, for the time being, in talk about their form of life rather than in the practical activity. In the middle of those with the hammers is a stick figure, who might be a sociologist, grinning happily, having mastered the interactional expertise while never having worked with hammer and anvil. As can be seen, the discourse of the smiling interactional-but-not-contributory expert is indistinguishable from the discourse of those with the hammers. That is the bold conjecture of the strong interactional hypothesis.

Origins of Interactional Expertise

The idea of interactional expertise has wide application. As well as being needed in some approximate form by successful participatory sociologists, ethnographers, and social anthropologists, mastery of interactional expertise is also the goal of specialist journalists; it is needed by salespersons and, as we will argue, by managers; it is often the medium of specialist peer review in funding agencies and journal editing where the reviewers are only sometimes contributors to the narrow specialism being evaluated; it is the medium of interchange within large-scale science projects, where again not everyone can be a contributor to everyone else's narrow specialism; it is, *a fortiori*, the medium of interchange in properly *interdisciplinary*, as opposed to multidisciplinary, research; it is, arguably, the basis of the assumption of false identities in Internet chatrooms;[23] finally, on those occasions when activists or other concerned persons are driven to it, it can be the medium of interchange between scientists and groups of the public.

Though it has very wide application, the idea of interactional expertise emerged for us as a result of our experience of sociological fieldwork, and this is how we will introduce it. Typically, sociologists who want to study areas of scientific knowledge that are new to them have to try to grasp something of the science itself. The sociologist begins with no specialist expertise—which is a level insufficient to do sociological analysis of scientific knowledge. The sociologist is likely to move rapidly through public understanding and primary source knowledge, which are also inadequate to allow for competent social analysis of scientific knowledge. With luck, however, interactional expertise, which does allow for social analysis of scientific knowledge, will eventually be attained.[24]

The transition to interactional expertise is accomplished, crucially, by engaging in conversation with the experts. Interactional expertise is slowly gained with more and more discussion of the science (or other

23. But confidence tricks of all kinds generally depend on a major contribution from the "mark"—the person being fooled (Maurer 1940). In the imitation game (see below) it is quite clear that the judge's job is to try to distinguish the true expert from the expert who has mastered only the language; the judge does not make the kind of contribution made by the mark.

24. In rare cases the sociologist might even progress to the level of contributory expertise. Contributory expertise can be attained only by practicing the science, and there is rarely an opportunity to do this if the sociologist has not undergone the full-scale training in professional institutions that are the prerequisite to certification. It is not impossible, however, in sciences which are not too difficult, as in Collins's contributions to parapsychological research, where he took an active part in designing and carrying out experiments (see Pamplin and Collins 1975). Also, some degree of contributory expertise may be attained where the science is not too far removed from the sociology, as in the case of artificial intelligence. Collins would claim that his books on artificial intelligence (1990; Collins and Kusch 1998), make contributions to the field itself as well as being sociological analyses. (Whether those contributions have been taken up is another matter.)

technical skill). Interactional expertise cannot always be attained—the science may simply be beyond the capacity of the analyst, as one of the authors, Collins, discovered when he attempted to do research in the field of the theory of amorphous semiconductors. After completing about thirteen hours of taped interviews with scientists, he had to concede that he could not understand enough of the science to reach a sufficient level of comprehension of the scientists' world to make any sociological headway. He had to give up. One characteristic of such a failure is that each new interview or discussion would start with a long and tedious period of explanation of how the science worked that repeated, approximately, the explanation that had marked the start of every other interview or discussion. All parties were equally bored by these explanatory sessions, and the interviews went little further.

In contrast, where interactional expertise is being acquired, there will be a progression from "interview" to "discussion" to "conversation" as more and more of the science is understood. There is no sudden "aha moment" that marks the switch to mastery of interactional expertise, but its steady acquisition can nevertheless be recognized. Above all, with interactional expertise, conversation about technical matters has a normal lively tone and neither party is bored. As things develop the day may arrive when, in response to a technical query, a respondent will reply "I had not thought about that," and pause before providing an answer to the sociologist's technical question. When this stage is reached, respondents will start to be happy to talk about the practice of their science and even give studied consideration to critical comments. Eventually respondents will become interested in what the analyst knows about the field because he or she will be able to convey the scientific thoughts and activities of others in a useful way. The sociologist who has just come from visiting scientist X may be able to tell scientist Y something of *the science* that X is doing or the kind of thinking that X is engaged in respect of some common problem. Sometimes the analyst will be able to introduce a new piece of science to a scientist.[25] Occasionally the analyst will be able to explain the scientific position of another party in a clearer way than the scientist him or herself currently understands it; this is because the analyst has heard the position explained at great length whereas the scientists

---

25. For example, in May 2005 Collins found himself explaining the "Christodoulou effect," a technical wrinkle in gravitational wave physics (Collins 2004a), to a gravitational wave physicist during the course of a workshop—the physicist had never heard of it.

may find communication with an academic rival difficult because of their different commitments and interests. By this stage, what were once "interviews" have become "conversations," not markedly dissimilar to the conversations the analyst will have with social scientist colleagues and, presumably, not that different to the conversations that one scientist will have with another. (For the sociologist, sitting in on the conversations that scientists are having with one another no longer seems like eavesdropping so much as participating.) In sociologist-scientist conversation of this kind *both* parties can speed things along by anticipating a technical point so that a longer explanation is avoided when existing mutual understanding is indicated by an interjection. In the same way the conversational partner's expression of a point may be helped, or a memory jogged, by a phrase which anticipates what is to come. At this stage jokes, irony, and leg-pulls are all recognized, and respondents will no longer be tempted to give a "pat" or formulaic answer drawn from a ready-made set of responses representing the canonical face of science. Mostly, respondents will talk to the analyst as they would talk to a colleague rather than an outsider, knowing that the standard recipe will not do. If, however, a respondent is encountered who does not know the analyst well, and is tempted to provide "officially approved" answers to questions, the analyst will have the skill to recognize the nature of the response and discount it or probe further; a sharp technical remark by the analyst can speedily change the whole tenor of such a conversation. As things go on the analyst may even develop the confidence to take a "devil's advocate" role in respect of some scientific controversy and argue a scientific case with which the respondent disagrees. The counter-case may be maintained well enough by the analyst to make the respondent think hard.

Where there is no developing interactional expertise, as in Collins's experience in the case of the theory of amorphous semiconductors, the conversations never become interesting to either party, the analyst can never transmit information, take a devil's advocate position or, crucially, distinguish between "pat" answers and real conversational interchange, nor between jokes and irony on the one hand and serious responses on the other. Worse still, though a field might be riven with controversy (as the theory of amorphous semiconductors was at the time of this fieldwork), the analyst cannot understand what the protagonists are disagreeing about, nor how deep the disagreements run, nor, with any certainty, who disagrees with whom! The contrast between these extremes—no expertise and a good level of interactional expertise—are very marked and quite unmistakable, at least by the fieldworker who has experienced both.

In spite of gaining very high levels of interactional expertise—to the extent of fulfilling a useful minor role in the transmission of scientific information among the scientists, or occasionally of giving a clear explanation to one party of the scientific position of another—the analyst is not going to be given a job or let loose in a scientific laboratory; that would demand contributory expertise. The analyst who has even the highest levels of interactional expertise may be able to *understand* scientific things, and to *discuss* scientific things, but is still not able to *do* scientific things.

The Parasitic Nature of Interactional Expertise

We can be fairly sure that a difference between interactional and contributory expertise is that contributory expertise is self-sustaining whereas interactional expertise is not. That is to say, a contributory expertise— such as gravitational wave physics—can be taught to new recruits and is passed on from generation to generation by apprenticeship and socialization; someone who has the contributory expertise can pass it to someone who does not have it. It is not at all clear that the same applies to interactional expertise. It is not at all clear that interactional expertise, which, in practical fields, is always interactional expertise *in another expertise*, can be passed from one person or generation to another (in the absence of contributory expertise). Interactional expertise in a specialism seems to be learned exclusively through interaction with communities who have contributory expertise in that specialism, not persons who have interactional expertise in that specialism. One would guess that, if the attempt were made to transmit interactional expertise in the absence of contributory expertise over several generations, it would rapidly become distorted as messages are distorted when they are passed on by word of mouth through many intermediaries. The point is that interactional expertise is skill in speaking a specialist language, and the nature of a whole language is a function of the whole environment, physical and social, in which it develops. Change the environment (e.g., remove the physical activity which is initially an integral part of the development of a language), and the language will change. But this does not mean that an individual immersed in the linguistic community cannot learn the language without being engaged physically with the physical world that gave rise to it, as we will argue at greater length in chapter 3.

The Relationship of the Specialist Expertises

This completes our initial five-step ladder of expertise. It starts from beer-mat knowledge, and goes to public understanding and primary source knowledge, all of which turn on ubiquitous expertises only. Then it makes the transition to expertises involving specialist tacit knowledge, the first of these steps being interactional expertise and the second being contributory expertise.

There is a transitive relationship between the five levels of the ladder. If you possess one of the higher levels you will possess, at least in principle, all of the lower levels but not *vice-versa*. There are, however, a few practical exceptions to the transitivity. First, as we will discuss in the next section, a contributory expert's interactional expertise may be "latent," i.e., not realized. Second, contributory experts may know the journal literature only at second hand rather than have the firsthand acquaintance of those whose knowledge extends only as far as the primary sources. Experts' knowledge does not come primarily from an exhaustive knowledge of the literature but from a familiarity with a subset of the literature, often at second hand, and always modulated by the opinions of other experts. Third, and for similar reasons, it may well be that specialists in general knowledge quizzes and the like could have a greater breadth of beer-mat knowledge than a domain specialist.

Hand-in-hand with the transitivity of the specialist expertises goes the transitivity of their pattern of distribution among the population. As we move up the scale from no specialist expertise, through beer-mat knowledge, popular understanding, primary source knowledge, interactional expertise, and contributory expertise, we find ourselves looking at smaller and smaller groups of people; the expertise becomes more and more esoteric. Popular understanding is limited to the numbers who read popular science books and articles in the science magazines and broadsheet newspapers. Once we get to primary source knowledge we encounter still smaller numbers, who tend to be driven by special health needs, local circumstances, or burning political agendas—forces which may also lead them to mix in the kinds of scientific circles where they are exposed to a deeper understanding of the issues.[26] Those with interactional

26. There are many examples in the literature in which it can be seen how the key citizen activists are driven by some combination of these motives and interests. For example, many of the case studies contained in Irwin and Wynne 1996 have this quality, as do the studies of Repetitive Strain Injury patients (Arksey 1998); AIDS treatment activists (Epstein 1995, 1996) and nuclear protestors (Welsh 2000).

expertise are fewer in number still, since gaining interactional expertise requires crossing social boundaries and spending a long time in alien social environments to which there is restricted access. Finally, those with contributory expertise may, in highly technical sciences, be limited to somewhere between a half-dozen and a few hundred. (Remember that all contributory experts are counted as possessing interactional expertise by definition—the numbers of interactional-but-not contributory experts are very small.)

Interactional Expertise and Interactive and Reflective Ability

Interactional expertise looks similar to but is distinct from other kinds of capacity that are part-and-parcel of the job of the sociologist, journalist, art critic, architect, and so forth. All these professionals need the ability to interact with other people, to talk smoothly about the domain which they have chosen to study or within which they exercise their judgment, to reflect upon their subject matter so as to articulate their findings or judgments, and sometimes to translate the expertise of one domain into the language of another insofar as this can be accomplished. These are capacities not necessarily shared by those with contributory expertise in the domain, and this raises a question about the transitivity of the relationship between contributory and interactional expertise. We claim that if one has contributory expertise in a domain one also has interactional expertise, but, if one does not have a ready ability to talk and reflect, then one is likely to have little in the way of interaction with others in respect of the expertise. As intimated, the resolution is to say that in the absence of the other kinds of capacities the interactional expertise of the contributory expert will be latent rather than expressed. What we mean by this is that, in order to realize the latent expertise, nothing new pertaining to the specific domain in question has to be learned. The things that have to be learned to realize the latent expertise are to do with the domain of talking, reflecting, translating, and so forth, not laser-building, or gravitational wave physics, or car driving, or whatever. That there must be a difference between latent interactional expertise and an absence of interactional expertise is easy to see: one could, at least in principle, reveal the interactional expertise of an inarticulate and unreflective contributory expert by skilled and persistent probing—from skilled interviewing one could learn something about the domain (this is what sociologists and journalists do in the case of inarticulate and unreflective respondents).

In contrast, *no amount* of probing will extract deep information about a domain from someone with neither contributory nor interactional expertise.[27]

We will give the labels "interactive ability" and "reflective ability" to the capacities that turn latent interactional expertise into expressed interactional expertise (these are the dispositions found in row 2 of the Periodic Table).

Interactive Ability

To repeat, possession of contributory expertise guarantees possession of at least *latent* interactional expertise. To realize the interactional expertise it is also necessary to possess interactive ability.

A lack of realized interactional expertise combined with a high level of contributory expertise is very typically exhibited by many fine artists who consider that their work must "speak for itself." "If the meaning of a painting could be expressed in words there would be no point in painting," as they might and do say. They make the point in practice by refusing to speak fluently about their work and by allowing their reflective discourse to atrophy.

On the other hand, as explained, in the role of art critic, journalist, sales representative, television or radio interviewer, and interpretative sociologist, the skills needed to interact with others are crucial. Without these skills the job cannot be done. In these roles, a high level of *interactive ability* is part of the *contributory expertise* pertaining to that particular specialism (though the specialism itself may be almost entirely devoted to gaining interactional expertise in other specialisms).

An important difference between interactional expertise and interactive ability is that the latter, unlike interactional expertise, is not parasitic—it can be passed from generation to generation. Interactive ability is, as we will call it, a "disposition," like kindness, or a loving nature, or a gift for observation, rather than a specialist skill. For example, parents who have the "gift of the gab" are likely to pass this ability on to their children. The point is that interpersonal skills are generalized abilities, not an expertise *in a special domain*. It is because interactional

27. The above explanation of how we use the term "latent" is a response to Evan Selinger who argues (personal communication) that you cannot have a latent expertise; a latent expertise is no expertise at all. Carolan (2006) also discusses these issues in the context of an interesting analysis of the role of interactional expertise in the development of farming skills.

expertise is *expertise in something* that it is unlikely that such an expertise could be passed on in the absence of continued contact with the "something." To repeat, one cannot imagine that interactional expertise would do anything other than die out if not refreshed from time-to-time by contact with those actually doing the thing—the contributory experts. It is the contributory experts not the interactional experts who define and develop the content of the language that the interactional expert tries to master.

Reflective Ability

Another generalized skill, which, with a little stretching, can also come under the heading of a disposition, is reflective ability. This, it is true, is a more professionalized and specialized ability than interactive ability because it is taught, quite self-consciously, in sociology and philosophy courses and the other critical disciplines. Like interactive ability, reflective ability is enormously useful in the building of interactional expertise. It is sometimes what makes the difference between the analyst and the specialist scientist when they are talking about an esoteric domain. Some scientists are actually proud of their lack of reflective instincts, boasting that "Philosophy of science is about as useful to scientists as ornithology is to birds."[28] This is perfectly correct, but it carries the corollary that it is not birds (scientists) who one should consult to learn *about* flight (science). Reflective ability, like interactive ability, can exist *sui generis* and be passed from generation to generation. Reflective ability is not reflective ability *in something*, it is just reflective ability. Reflective ability is, again, part of the contributory expertise of the social analyst of science, the art critic, and so forth.

We can assemble some of these relationships in diagrammatic form. The large circles in figure 3 represent respectively the *concepts* of contributory expertise, interactional expertise, and interpersonal and reflective ability (combined for simplicity's sake). If, as we say, contributory and interactional expertise are related transitively, then spaces 1 and 5 are empty of people: those who possess contributory expertise will also possess interactional expertise, either latent (located in space 4), or realized (located in space 7). In practice anyone who has acquired interactional expertise without acquiring contributory expertise is likely to be a member of one of those professions that turn on interactive ability and will

---

28. Commonly attributed to Richard Feynman.

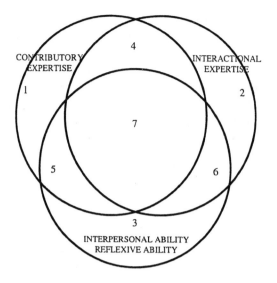

Figure 3. Relationships among some expertises

therefore be found in space 6 rather than space 2. Thus, space 2 is likely to be empirically empty if not quite as *logically* empty as spaces 1 and 5. Space 7 contains those few social analysts who are also technically competent as well as articulate scientists and technologists. Space 3 is occupied, among others, by philosophers of science and sociologists of science (as opposed to sociologists of scientific knowledge), who do not need interactional expertise for their style of work. In practice that group is likely to stress reflective ability rather than interactive ability.

Acquiring Expertise:
Five Kinds of Face-to-Face Knowledge Transfer

What is involved in the acquisition of an expertise under our model? We take as our example the ability to build a working model of a piece of novel scientific apparatus. We ask the following question: Suppose scientist B from B-lab wants to learn to build one of those heavy and nontransportable A-mators so far only built successfully by scientists in A-lab. We know from previous studies that, barring reinvention "from scratch," the right thing for scientist B to do is visit A-lab for a period and spend time talking with the A group as they work on the A-mator. Alternatively, a group from A-lab can visit B-lab for a period and hold extensive discussions. We then ask what scientist B learns from a visit.[29] Five problems related to knowledge transfer can be identified that could be resolved by extensive meetings between scientists from different laboratories:

1. *Concealed Knowledge:* A does not want to tell "the tricks of the trade" to others, or journals provide insufficient space to include such details. A laboratory visit can reveal these things but only if it is in the direction B to A.

2. *Mismatched Salience:* There are an indefinite number of potentially important variables in a new and difficult experiment, and the two parties focus on different ones. Thus, A does not realize that B needs to be told to do things in certain ways, and B does not know the right questions to ask. The problem is resolved when B watches A work so, again, a B to A visit is needed.

3. *Ostensive Knowledge:* Words, diagrams, or photographs cannot convey information that can be understood only by direct pointing, or demonstrating, or feeling. This can be accomplished by B visiting A-lab.

4. *Unrecognized Knowledge:* A performs aspects of an experiment a certain way without realizing their importance; B will pick up the same habit during a visit, while neither party realizes that anything important has been passed on. Much unrecognized knowledge becomes recognized and explained as a field of science becomes better understood, but this is not necessary.[30] Again, the B to A direction is best.

5. *Uncognized/uncognizable Knowledge:* Humans do things such as speak acceptably formed sentences in their native language without knowing how they do it. There are experimental counterparts to such abilities. They are passed on only through apprenticeship and unconscious emulation. At the end of the transfer neither party can describe what has been transferred; they may not even notice that anything has been transferred. Insofar as the emulation is of practices, it will be necessary, once more, for B to visit A.[31]

Concealed knowledge is to do with lies and secrecy, not the nature of knowledge transfer, so it is not central to this discussion. Turning to the other four problems, we have discussed them in terms of the transmis-

29. The elements of this discussion are originally worked out in Collins 1974 (or see 1992), and 2001a (or see 2004a, chap. 35).

30. For a discussion of the problem with reference to Italian violins, see Gough 2000.

31. For the first use of this list, see Collins 2001b.

sion of contributory expertise. But what if it was interactional expertise alone that was to be transferred? Would an A-group's visit to B-lab be as good as B's visit to A-lab? How important to language acquisition is it that the talk be conducted in the presence of the apparatus? In chapter 3 we will return to the problem, but it may be that we already know the answer. The point is that none of those things that A thinks are salient and B does not realize are salient (2), and none of the things that A can only point to (3), and none the things that A does without realizing it (4), and none of the things that no-one can articulate (5), figure in A's *language*. Therefore, if B is interested in acquiring only A's language, then B has no need to acquire the understandings that don't have any counterpart in the language. Therefore, at a first approximation, a visit of an A-group to B-lab should be just as useful for acquiring interactional expertise as a visit of B to A-lab.[32] This is but a first approximation, since it is probably the case that a visit of a single individual to B-lab is not going to be a very good way of transferring interactional expertise—this would be like our trying to learn fluent French from the visit to our house of a single French person; what would be needed would be extended visits of a group of scientists from A-lab to B-lab.[33]

If it is true that, in principle, so little in the way of knowledge is necessary for the acquisition of interactional expertise that it can be acquired without being in the presence of the apparatus, then interactional expertise is very limited. But that only makes it the more interesting, and, at first glance, counter-commonsensical, that interactional expertise is so important in science and technology. Remember, if, as per the strong interactional hypothesis, a person with contributory expertise cannot say anything in virtue of having that expertise that a person with interactional expertise cannot say, then interactional expertise is just as good in forums that work through the medium of language as contributory expertise.

It now seems a shame that the analysis of the immediately preceding paragraphs has not been tested when studies of scientific knowledge transfer have been carried out. For example, in an early study of the trans-

---

32. Even though this is a "first approximation," it is true only in principle. In practice, having an apparatus present is likely to make the conversation easier, even if it is only interactional expertise that is at stake. The point remains, and it will be discussed at length in chapter 3, that where circumstances do not permit proximity to the apparatus or its equivalent, interactional expertise can be acquired even though it will need extra effort.

33. We are grateful to an anonymous reviewer of the manuscript for pointing this out to us.

fer of laser-building ability no attention was paid to how much of what was being learned by one scientist in another scientist's laboratory was learned by talk and how much by watching and doing.[34] There would be very many confounding factors in a naturally occurring situation, but perhaps this kind of question could be the subject of an experimental study.

The same oversight can be found in a 1990 discussion by the same author. Collins asked how a spy, who pretends to be a native of a town he has never visited—Semipalatinsk—would be caught out in conversation with a native of Semipalatinsk. He claimed there were two ways for a spy who had never visited the town to learn about Semipalatinsk: learning from books, photographs, and so forth, and learning through conversation with an expatriate native of the town. He claimed that, in contrast, genuine natives would also learn from experiencing the physical reality of the town and from immersion in the linguistic culture. Collins argued that the spy would fail a "Turing test" when compared with a native if the interrogator was also a native because of the things natives could say to each other as a result of the second two kinds of experience but which the spy would not be able to say.

We can now see that Collins failed to consider the extent to which the spy could have been trained to pass the interrogation by immersion in the language of Semipalatinsk alone if, say, that immersion was arranged by his spending a year with a group of exiled Semipalatinskians. Such a group could have transferred the capacity to handle nuances of pronunciation or vernacular speech and all that can be acquired through the development of interactional expertise. On the arguments set out in respect of the transfer of scientific knowledge, physical experience of the town would not be important since everything that could be said as a result of physical immersion could be said by the expatriates. The spy, then, could have mastered the interactional expertise of Semipalatinski-anism to the level that the expatriates still possessed it. The spy interrogation test, which is very like an imitation game (see below), is a test of interactional expertise not contributory expertise.

As it happens, Collins and Kusch, in their 1998 book, did better (though without realizing it at the time). They discuss the role of Cyrano in the play *Cyrano de Bergerac* by Edmund Rostand. Cyrano agrees to write love letters on behalf of another suitor. They write:

34. The study in question is Collins 1974 (or see 1992).

. . . suppose Cyrano had never known love but, notwithstanding, had the skill to write the prose? We may imagine he had read the relevant literature and poetry and had frequented the society of those who did know love. In other words, Cyrano would have understood the institution of love and love letters even though he had never felt the individual emotion himself. (94)

In retrospect we can now describe this as an instance of the acquisition of the interactional expertise of love in the absence of the contributory expertise of love and an illustration of the individual embodiment thesis (to be discussed in chapter 3).

CHAPTER TWO

The Periodic Table of Expertises 2: Meta-expertises and Meta-criteria

We now turn to the lower section of the Periodic Table. Meta-expertises are expertises used to judge other expertises. There are two kinds of meta-expertise: *external meta-expertise*, which does not turn on acquisition of the expertise itself, and *internal meta-expertise*, which does involve an acquaintance with the substance of the expertise being judged.

External Judgment: Ubiquitous Discrimination

We can make a start on defining the class of those who are in a position to judge experts and expertises by noting that in respect of some kinds of judgment the boundaries extend to the very edge of the general public. Just as there is ubiquitous expertise, there is also ubiquitous meta-expertise, which we will call "ubiquitous discrimination." In such a case the Problem of Extension dissolves, as with other ubiquitous expertises. Ubiquitous discrimination, like other ubiquitous expertises, is acquired as part-and-parcel of living in our society.

For example, those with little scientific knowledge can sometimes make what amounts to a *technical* judgment on the basis of their *social* understanding. The judgment turns on whether the author of a scientific claim appears to have the appropriate scientific demeanor and/or the appropriate location within the social networks of scientists and/or not too much in the way of a political and financial interest in the claim.

Ubiquitous discrimination is what we have all been learning since we could speak, and it is just a particular application of our regular judgments about friends, acquaintances, neighbors, relations, politicians, salespersons, and strangers, applied to science and scientists within Western

Table 2: The lower part of the periodic table of expertises

META-EXPERTISES	EXTERNAL (Transmuted expertises)		INTERNAL (Non-transmuted expertises)		
	Ubiquitous discrimination	Local discrimination	Technical connoisseurship	Downward discrimination	Referred expertise
META-CRITERIA	Credentials		Experience		Track record

scientific society. For example, many members of such a society, just by being members, are able to discriminate between what counts as science and what does not. This is the ubiquitous judgment on which we rely when we dismiss certain beliefs such as astrology from the list of contributors to the *scientific* element in technical decision-making. Most members of our society have sufficient social understanding to know that the standards and social and cognitive networks of astrologers do not overlap with the standards and social and cognitive networks of scientists.[1]

Another illustration of this point emerges from the dispute over the Moon landings. There is a view that the Moon landings were faked by the Americans, the events being filmed somewhere in a desert in the Western USA.[2] The groups who believe this cite various anomalies in the films, such as the quality of the shadows, the way the flag flapped, or some such. Technically we are in no position to judge whether the films were real; certainly it would be quite possible with the film technology of the time to have faked the whole thing. If we turn to those with more than the average level of technical expertise, we find that even there fakery of this sort is not easy to rule out. For example, it is reported that some American astronauts and other technical people disputed the filmed evidence of the first Russian "space walk" (EVA) in 1965. The astronaut David Scott reports his reaction at the time as follows:

> I was on my feet pacing by this point. "If this EVA is real, they're not only ahead but pretty far ahead, at that. What proof do we have that this guy really went outside?"

1. Poor social judgments are the problem with those who believe in, say, newspaper astrology *as a scientific theory.* They are making a social mistake—they do not know the locations in our society in which trustworthy expertise in respect of the influence of the stars and planets on our lives is to be found. In chapter 5 we will develop these ideas further using Wittgenstein's notion of "family resemblance" as the basis of our theory of what should count as a fringe science and what sciences should count as continuous with Western science.

2. Susan Carter brought the relevance of the Moon landings to our attention.

The first grainy photographs released to the world press of Alexei Leonov floating in space sparked a heated debate in the West. Some claimed the photos were faked. They simply would not accept that the Russians had chalked up another first.[3]

Given astronauts' ability to doubt the Russian space walk, there is nothing technical to stop ordinary people from doubting the story of the Moon landings.

What stops us doubting their validity is, once more, ubiquitous discrimination, which is a social judgment. It is beyond the bounds of sociological credibility, even ordinary people's sociological credibility, that the thousands of people involved in the Moon missions could all have been organized to lie so constantly and consistently; we know that if there were any possible credence to the story of the fake, the Russians, deeply involved in the cold war as they were, would have exploited the doubts—yet they did not. Our ubiquitous social discrimination allows us to be sure about the Moon landings, even while the technical discrimination of those who were fairly close to the events does not.

To see how this kind of discrimination works in a more difficult case, consider cold fusion. As the cold fusion saga drew to a public conclusion around the last years of the twentieth century, most reasonably literate members of Western society, who knew nothing of cold fusion beyond what they had seen on the news or read in the newspapers, "knew" that cold-fusion had been tried and found wanting. Though there was a time when cold fusion was contiguous with science as we knew it, it was now understood that its cognitive and social networks no longer overlapped with those of legitimate scientific society. This knowledge had nothing to do with scientific competence. On the contrary, it was vital to ignore scientific credentials, and even track records of success, if a socially appropriate judgment was to be made. Thus Martin Fleischman, the co-founder of the cold fusion field, had an enviable track record for success in the sciences, was immensely well qualified, was honored as a Fellow of the Royal Society, and had both interactional and contributory expertise in cold fusion, yet still believed in the effect, contrary to the scientific consensus. What people in Western societies had in common is what they had heard about cold fusion in the broadcast media (popular understanding). Their consensual view, insofar as they had one, emerged from making *social* judgments about *who* ought to be agreed with, not

3. Scott and Leonov 2004, 124.

scientific judgments about *what* ought to be believed. To expect the citizen to be sufficiently educated in science as to be able to make scientific judgments in disputed esoteric matters like this was the futile aim of the old COPUS-like organizations. (Because "distance lends enchantment" they will, however, find many who will enthusiastically endorse their aims—ordinary citizens who think that having read a popular book or two they now understand cold fusion better than a Fellow of the Royal Society with a lifetime of relevant practical experience.)[4] The crucial judgment, however, is to "know" when the mainstream community of scientists has reached a level of consensus that, *for all practical purposes,* cannot be gainsaid in spite of the determined opposition of a group of experienced scientists who know far more about the science than the person making the judgment. Note that this is not the sort of judgment that we would expect even an immaculately qualified scientist from "another planet" to be able to make. A scientist from another planet, reading published papers for and against cold fusion, would have difficulty working out who was right; the scientifically ignorant citizens of this planet, in contrast, had a relatively easy decision to make.[5]

External Judgment: Local Discrimination

There is a quite different version of discrimination that pertains to more specialist groups. At this point we need to introduce one of the most influential studies of the nature of expertise—Brian Wynne's examination of events on the Cumbrian fells following the Chernobyl nuclear meltdown in 1986. Wynne's work has been influential in both a positive and negative way. It has helped to establish the idea that technical expertise can be found beyond the normally recognized qualified groups, but it has also given rise to much confused thinking about the extent to which laypersons can be experts. Wynne looked at the interaction between the UK Ministry of Agriculture Food and Fisheries (MAFF) scientists and the Cumbrian sheep farmers after radioactive fallout contaminated their pastures.[6] Wynne argued that the expertise of the sheep farmers in respect of sheep ecology should not have been ignored by the scientists—the sheep farmers had what he called "lay expertise." Though spotting the expertise

---

4. Which is the flaw in the "deficit model" of public understanding of science.

5. See Collins 1999 for a similar argument in respect of the rejection of claims about the existence of gravitational waves.

6. Wynne 1989, 1996a, 1996b.

of the sheep farmers was insightful, "lay expertise" was an unfortunate choice of term because of its potential to cause confusion. For example, the term has often been interpreted as meaning that laypeople possess specialist expertise. It would have been better if Wynne had talked of experts without formal qualifications. For example, the sheep farmers are not laypersons; they are experts in sheep farming who happen to have no paper qualifications. The sheep farmers have a specialist contributory expertise. Their expertise is esoteric, highly relevant to the ecology of sheep on radioactive fells, but, unfortunately, it went unrecognized by Ministry scientists.[7] As a group of contributory experts, discussion of the sheep farmers' expertise belongs in chapter 1.

But Wynne also found that in addition to contributory expertise the farmers possessed what we will call "local discrimination," which is more properly a topic for this chapter. Soon after the Second World War, the Windscale-Sellafield nuclear processing plant was built on the Cumbrian fells, and the farmers thus had long experience of the nuclear industry's pronouncements concerning radioactive contamination; they knew that these pronouncements could not be taken at face value. An outsider, with less experience of discussions of radioactive contamination in this particular social and geographical location, would not have been able to judge the pronouncements with such finesse. The sheep farmers were able to discount statements by the nuclear industry's spokespersons, but in this case it was a result of local experience rather then a more generalized discriminatory ability developed over a lifetime of social and political education. Local discrimination in such a case is analytically, and sometimes practically, distinct from contributory expertise. For example those long-term residents of the area surrounding Windscale, who knew nothing about sheep farming, almost certainly knew quite a lot about the nature of the assurances offered by the representatives of the local nuclear industry.

Another study by Wynne reinforces both the point and the confusion between *local* and *ubiquitous discrimination*. Wynne describes the experience of apprentices working in the radioactive materials industry. He suggests that the apprentices felt they had no need to contribute to their own safety by trying to understand the science of radioactivity because they were "intuitively competent sociologists" and "vigilant and active seekers of knowledge . . . tacitly and intuitively, positioning themselves, using their knowledge of their social relationships and institutions."[8] Wynne

7. Or so we understand.
8. Wynne 1992, 39.

argues that the apprentices used their social understanding as a basis for trust in their employers. In a later article, referring to the same group, he says that these apprentices' "technical ignorance was a function of social intelligence."[9]

There are two ways of looking at Wynne's discussion of the apprentices. It could be an example of *local discrimination*. In that case the apprentices would be seen as using their local competence in understanding the trustworthiness of their particular employers and their own place within the social networks of trust operating in that particular workplace to assess the safety of the procedures in which they were involved. On the other hand, the apprentices might be doing no more than any of us do when we trust the institutions which surround us. For example, when we put money in the bank we do not say that we have no need to understand economics because we are "intuitively competent sociologists" and "vigilant and active seekers of knowledge . . . tacitly and intuitively, positioning ourselves, using our knowledge of [our] social relationships and institutions." This language seems quite unnecessarily folksy and romantic. Would we say we are using our social understanding as a basis of trust in the bankers and that our economic ignorance was a function of social intelligence? Well, yes, but it is a passive kind of knowledge—something that we learn in the same way as we learn a language. It is the kind of "knowledge" that is the basis of all trust in society. It will differ from society to society, and it will occasionally be breached (as when there is a "run on the banks"). It would be, nevertheless, a ubiquitous expertise and as such would not really solve any local and technical problem. What should be clear is that for discrimination to be "esoteric" it must be local discrimination not ubiquitous discrimination. In the case of the apprentices, if their discrimination was of the ubiquitous kind, then no amount of sugaring the pill makes them other than uninformed in respect of their own safety; if it was local discrimination—they had learned who to trust and who not to trust as a result of long experience in their particular workplace—then that is a different matter.[10]

9. Wynne 1993, 328.

10. Gravitational wave scientists report that they use the following criteria to judge whether an experiment by another scientist needs to be taken seriously: faith in experimental capabilities and honesty, based on a previous working partnership; personality and intelligence of experimenters; reputation of running a huge lab; whether or not the scientist worked in industry or academia; previous history of failures; "inside information"; style and presentation of results; psychological approach to experiment; size and prestige of university of origin; integration into various scientific networks; nationality.

Problems of External Discrimination

Local discrimination, like ubiquitous discrimination, is a meta-expertise that is *external* to the expertise being judged because it does not depend on the understanding of the *expertise* being judged but upon an understanding of the *experts*. It is, as explained at the beginning of this chapter, a way of reaching technical conclusions via nontechnical means. In general it is very unreliable because of the temptation to read too much into stereotypical appearances and stereotyped behavior. It was this tendency to read too much into appearance that was exploited by the "scientists in white coats" who, for many years, assumed and were given a license to speak with authority on almost any subject. It could be said that the stereotype of the scientist was what gave rise to the misleading picture of the power of logical thought and experimental genius. It could well be argued that the public's misunderstanding of the MMR controversy (see chapter 5) was partly a product of the demeanor of Andrew Wakefield—a young, handsome, kindly and, seemingly caring doctor up against the establishment. Again, the remarkable Dr Fox Lecture should warn us against judgments based on demeanor. Dr Fox was an actor hired to present a lecture to university audiences. The content of his talks were impressively technical-sounding gobbledygook. Answering a questionnaire after the performance, large proportions of the audience expressed themselves well satisfied with what they had heard.[11] Of course, there is also an equally misleading counterstereotype: the mad, monster-creating scientist which gave rise, at the time of the controversy over genetically modified organisms, to the famous British newspaper headline referring to "Frankenstein Foods." Nevertheless, judgments on the basis of demeanor and social position are often made within science (see footnote 10), so we should not dismiss them out of hand when they are made by the public. Furthermore, there are restricted and special circumstances where public discrimination seems sound. Some examples have been provided above: the Moon landings, cold fusion, and circumstances where members of the public were able to induce from past misleading statements that current statements about safety and the like were also likely to be misleading or, perhaps, vice-versa in the case of Wynne's apprentices. It should be noted that these examples turn on either a fairly good understanding

---

The first, second, sixth, eighth and tenth items could be said to be matters of local discrimination with the remaining items matters of ubiquitous discrimination.

11. Naftulin, Ware, and Donnelly 1973.

of social life or enough relevant experience to give rise to fair inductive inferences and do not imply the reliability of less well-founded discriminations. When the conditions are met, we seem to have what we might call "transmuted expertise"—a transmutation of social knowledge into technical knowledge. What we can say for certain, a point we will return to in the final chapter, is that discrimination of this kind is not part of the *legitimate* methods of science. When scientists discriminate in these ways, they do not trumpet the method in publications nor in any other part of the "constitutive forum."[12] Likewise, when the method is used by the public, it does not, and should not, be accepted as a legitimate input to scientific method.

Internal Judgments of Expertise and Their Problems

What other ways are there of judging between experts? The standard method of choice is by reference to the qualifications of the expert.[13] This we now know to be inadequate because it is possible to have expertise, and that includes specialist expertise, in the absence of qualifications. Thus, Wynne found that the uncredentialed Cumbrian sheep farmers knew a great deal about the ecology of sheep, the prevailing winds, and the behavior of rainwater on the fells that was relevant to the discussion of how sheep should be kept so as to monitor and reduce the impact of the radioactive fallout.

A still better documented case of uncredentialed persons gaining expertise is Epstein's study of AIDS treatment activists in the 1980s.[14] In 1985 a new drug, AZT, was about to be subjected to double-blind randomized control trials. AIDS sufferers were concerned that too many of those who were assigned to the placebo groups would die before the drug was approved. They therefore began a campaign for the introduction of speedier testing regimes, the relaxation of test protocols, and earlier release of potentially beneficial treatments. At first the suggestions of the activists—a group whose members' dress codes and presentation-of-self was as far from the world of medical orthodoxy as it was possible to be—were resisted. Robert Gallo, the co-discover of HIV, is reported as being initially hostile to the AIDS activists, saying to members of one of

12. Collins and Pinch 1979.

13. But bear in mind that even the most well-trained and accredited professional sometimes turn out to be incompetent.

14. Epstein 1996. See also Collins and Pinch 1998, or 2005 for a summary of Epstein's study.

their pressure groups, "the AIDS Coalition to Unleash Power" or ACT UP: "I don't care if you call it ACT UP, ACT OUT or ACT DOWN, you definitely don't have a scientific understanding of things" (Epstein 1996, 116). The activists, however, undertook an arduous course of self-education and learned the language of medical discourse. Most importantly, they added to a developing understanding of microbiology and statistics their experience of how AIDS sufferers would actually respond to the demands placed upon them by the protocols of randomized control trials. They knew that these demands were unrealistic: since death was in constant prospect, the groups regularly smuggled untested cures from Mexico, continued to take other drugs which were banned and which had the potential to confound test statistics, and even shared placebos and trial drugs between experimental groups. ACT UP knew that the randomized control trials were not working as the scientists assumed they were.

Eventually, the activists gained so much interactional expertise in research design that, allied with their experience, they were able to make real contributions to the science that were warmly embraced by the scientists. Gallo was to come to say of one of their leaders that he was "one of the most impressive persons I've ever met in my life, bar none, in any field. . . . I'm not the only one around here who's said we could use him in the labs." Gallo is also said to have described some activists' scientific knowledge as of an "unbelievably high" standard. He said: "It's frightening sometimes how much they know and how smart some of them are" (Epstein 1996, 338). The AIDS activists, though unqualified in any field that bordered on medical science, eventually trained themselves to a point beyond that of the Cumbrian sheep farmers, a point at which the science community took them very seriously indeed, not least because it enabled them to do better science.

Any criterion of expertise has to allow groups such as the Cumbrian sheep farmers or the AIDS activists to be included in the category of expert, and that is why the criterion of formal qualification or accreditation is too exclusive. We suggest that a more important criterion than qualifications is *experience*. If there is to be a general criterion of expertise, experience is the leading candidate. The criterion of experience would include the Cumbrian sheep farmers, the AIDS activists, and the like.

Note that there is no problem about judging expertise at the lower levels of the ladder. A general knowledge quiz such as Trivial Pursuit can discriminate adequately between levels of beer-mat knowledge, while higher level quizzes or examinations can discriminate proficiency in popular understanding or primary source knowledge. Indeed, much of our education system is dedicated to discriminating precisely at these levels, and hence the perennial complaints of employers that graduates come to them unfit for the workplace, where a different kind of expertise—the level of expertise associated with doing rather than knowing—is required. It is at these higher levels of expertise that the problems of judging expertise, both practical and conceptual, arise. We begin our investigation of the deeper problem by looking at those who pass themselves off as experts whereas according to most criteria they are not. We start, then, with hoaxers, frauds, and confidence tricksters.

Hoaxers, Frauds, and Confidence Tricksters

There are bogus doctors, bogus lawyers, bogus nurses, and bogus paramedics, bogus gas and electricity meter readers, and bogus traffic police. There has been at least one bogus Oxford don, at least one bogus army general—who turned out to be a woman posing as a man—a bogus Roman Catholic priest, and bogus CIA men. It seems, then, given the right conditions, people are ready to attribute almost unlimited expertise and authority to people whose behavior can, given a stretch of the imagination, be interpreted as indicating that they have it.

The phenomenon has been marvelously exploited in fiction. To pick one example among many, Jerzy Kosinski's book, *Being There* (1971) subsequently made into a film of the same name, shows an educationally subnormal, sexually neuter, but well-dressed gardener, the eponymous "Chauncey Gardiner," rise from unemployment to become president of the United States complete with a reputation for innovatory sexual adventure. It happens because a band of hangers-on reinterpret his minimal vocabulary as the profound and gnostic discourse of a sage and rake.

In relational theories of expertise there is a problem about how to deal with frauds and hoaxes. If expertise is attributed to a hoaxer, there is little more to say about it—the relevant topological location in the network has been achieved. The problem for a relational or attributional theory is what is special about hoaxes and frauds (before they are exposed), as opposed to genuine exercises of expertise. Indeed, this problem has been made into a virtue—the study of the attribution of the label "hoax" being the very point of some analyses.[15] Since we are concerned with judgment of expertises, however, we need to ask which kinds of role are more or less difficult to fake: Which kind of expertises is it easy to make a mis-

---

15. See, for example, Brannigan 1981.

take about, which are more easy to judge, and why? Thus, in the case of the "expertise" of, say, lying in bed in the morning, there is no expertise to fake, so anyone can say they are an expert in it without fear of contradiction: there are no confidence tricksters when it comes to lying in bed. There are, on the other hand, few or no confidence tricksters when it comes to solo violin-playing. At least, there are no confidence tricks in which a solo performance with an orchestra playing a well-known piece is part of the scam. In between are all the interesting cases, some more easy to fake than others.

A famous case is that of the trivially simple computer program, ELIZA, which was easily able to fake the expertise of a Rogerian psychotherapist. More recently, and uncomfortably "close to home" for some of the readers of this book, there is the case of the Sokal hoax. Alan Sokal submitted a manuscript for publication in the journal *Social Text* using the stylistic clichés of the semiotic turn in the cultural analysis of science; the journal published the paper only to have it revealed as a hoax, Sokal proclaiming "the emperor has no clothes." The idea of a hoax of this type is all in the revelation; if the perpetrator can show it is easy to pretend to have the expertise in question, then the expertise is made to look more like lying in bed or Rogerian psychotherapy than solo violin-playing. Sokal's hoax may have exposed lax editorial practices at *Social Text*, but it reveals little more since hoaxes are not so hard to pull off even in theoretical physics.[16]

16. To be exact, ELIZA's mistakes tended to be in its language-handling rather than the substantive content of its output. For discussion of this case, see Weizenbaum 1976; Collins 1990. Sokal's more extended conclusion, and that of many of his admirers, namely there is a large gulf between the integrity of the social and the natural sciences, was not borne out by events. Not much later a number of papers by the Bogdanov brothers on string theory were published in a variety of physics journals, and a long argument followed about whether they were genuine or a hoax; not being to pin down whether a paper really is a hoax after extended examination is, perhaps, still more embarrassing than the hoax carried off by Sokal. It is likely that at the cutting edge of all disciplines there are areas where no-one is really sure about what the new conventions should be, just as in the case of the avant-garde in the arts.

The original Sokal hoax is Sokal 1996. For more references see http://www.physics.nyu.edu/faculty/sokal/#papers. For the Bogdanov brothers event see http://math.ucr.edu/home/baez/bogdanov.html.

Note that there is a difficulty for those who would want to defend a journal with a "postmodernist mission," such as *Social Text*, from the predations of hoaxers such as Sokal. The difficulty is that, even to accept that Sokal has transgressed a boundary, they have also to accept that there is a difference between the genuine exercise of an expertise and its attribution. They have to agree, then, that even if there is nothing but attribution to everyone else's expertise, there is something real about their own.

Hoaxes and frauds are more easy to carry off than they should be because of the well-known tendency of their targets to "repair" deficiencies in the skill of the perpetrator, especially when they have something to gain by believing in the performance.[17] Part of the skill of the professional con-artist is to make the victim believe that he or she can bring them great financial gains, but the principle is universal: in nearly every case of a bogus performance, life for those around the fake will be much more inconvenient if they have carry out a complex investigation and perhaps replace a hitherto trusted colleague with someone else. That is why, even in the case of solo violin-playing, we have to say that the musical piece must be a "standard." If it is not a standard, then the audience, who have paid good time and money to hear a virtuoso performance, will be all too ready to believe that that is what they are hearing; they might, for example, believe they are present at a rendition of some avant-garde composition, or a piece of "conceptual art" the nonmelodiousness of which is the very point, since it asks questions about the meaning of music.

The role of avant-garde artist has been nicely satirized in the 1961 comedy film, *The Rebel*. In the film the comedian Tony Hancock plays an incompetent and untrained "artist" who finds himself sharing a garret-flat in the Bohemian quarter of Paris in the 1950s. As a result of a series of accidents, he is "taken up" by Bohemian society, and for a while his smears are treated as great works of art.[18] We are amused but not sur-

---

17. Collins and Pinch (2005, chapter 1), discuss fakes in general and a case study of bogus doctors carried out by Collins and Hartland. Bogus doctors are rarely unmasked as a result of medical mistakes because the surrounding team is ready to "cover" for them. Novice doctors, even though they have been through medical schools, have no experience and are expected to be unfamiliar with hospital conventions. As a result many bogus doctors have the opportunity to learn on the job and become quite successful. Experience, what we consider to be the most reliable criterion of expertise, stands up well when confronted with the bogus doctor case. A bogus doctor who survives long in the profession becomes a skilled doctor, even though he or she may have no certificates. As a result medical professionals who work with bogus doctors can be astonished when they discover that they have been fooled:

> I've never been so shattered in my life when a nurse came up to me and said the CID [the detective branch of the police] had been there . . . and I said, "What for?" and they said, [Carter] "had never qualified." I felt as if I had been hit with an atom bomb. . . .
>
> Had we been asked, and this was the general opinion of everybody when this came out, had we been asked to pick a doctor who was bogus, he would have been the very last of them all. (Quoted in Collins and Pinch 2005, 47)

18. *The Rebel* (*Call Me a Genius* in the USA) seems at first sight to be an exact parallel to *Being There*, but the difference is that there are kinds of painting that cannot be faked

prised at this outcome because we know that with the avant-garde, as the very name implies, there are no established conventions of artistic practice upon which to base a judgment.[19] That is also why we, or at least some of us, are relieved when we learn that Picasso was a brilliant artist within the conventions of realist depiction before he began to push forward the frontiers of art; knowing that he was so talented in ways which are relatively easy to judge, we can feel more secure in our appreciation of his less conventional works. In the same way we might like to know whether well-known avant-garde artists such as Tracy Emin and Damien Hirst can draw well: if we knew that, we could use their more easily understood skills as a proxy for their talent in a world where conventions provide no scaffolding for judgment.[20]

Technical Connoisseurship

The exercise of expertise within an established convention is, of course, convention-bound. It is not that, say, realist painting contains within itself some universal standard obvious to all: the conventions of what we understand to be realism in art have had to be established and change

---

without artistic talent (successful art forgers must be technically accomplished), whereas that may not be true of the presidency because of the way the president is embedded in a body of advisors.

19. The notion of the avant-garde helps to explain how bogus doctors maintain their presence in medical settings before they have learned on the job. Novice doctors are drawn from training hospitals in many different countries, and medical practice is sufficiently open to allow for a degree of variation which, while it may not be so great as in art, still provides a lot of leeway.

20. When all the virtuosity seems to lie in establishing a new convention rather than in executing a skill, art seems to be reduced to marketing, and it is not unreasonable to feel uncomfortable. Thus, we do well to be concerned that the Saatchi brothers, who run what was the UK's most successful advertising agency, are also the country's most successful collectors of avant-garde art and, co-extensively, the most powerful definers of its value. Obviously there is healthy ground for concern about this relationship if one believes that art is more than advertising.

This is also the easily understood and very reasonable reaction of critics to modern studies of science which take themselves too seriously as epistemology rather than methodology. In sociology of scientific knowledge, much of what used to be taken to be the exercise of skills within a convention is now understood to be coextensive with the establishment of a convention; this, for example, is the consequence of the experimenter's regress for the use of replicated experiments in the establishment of the existence new phenomena—those experiments that are counted as well executed are those experiments that produce what are taken to be valid findings under the new conventions of seeing (Collins 1992). The hostile reaction is quite unreasonable to, say, sociology of scientific knowledge as a methodology, however.

from epoch to epoch.[21] Nevertheless, *within a stable convention* virtuosity can be recognized.

The ability to recognize skillful practice itself improves through practice and that this is the case is recognized in terms such as "connoisseurship."[22] Connoisseurship is a meta-expertise. A connoisseur is, according the Chamber's Dictionary, "a person with a well-informed knowledge and appreciation." The dictionary definition tells us that the knowledge and appreciation are usually applied to fine food, wine, or the arts. But connoisseurship—that is, judgment honed by exercise—can be applied to all expertises, and we describe it when applied to such other expertises as "technical connoisseurship."

Consider, for example, that a contractor is employed to make major alterations to a house. At various stages, and particularly at the end of the job, it has to be decided whether the work has been finished satisfactorily. Imagine that some tiling has been done in the new bathroom. How even should the tiling be? How clean and square should the grout lines be? When has the job been finished? One can see immediately that there are conventions that give meaning to bathroom tiling, conventions that would be unknown to someone who had never lived in a society with tiled bathrooms. Some of the conventions can be set out as formal standards. For example, to go around curves, should the tiles be cut with a diamond saw, or cut square to an approximation to the curve, or chipped and hacked, the gaps being filled with grout? Some-

21. Difficulty of execution of a skill is, in the last resort, independent of convention. For example, suppose I decide to express my private artistry by peeling an apple in a spiral such as to produce a long unbroken ribbon of apple skin. To make a very long unbroken ribbon (imagine it only a couple of millimeters wide), might take months and months of practice, but there is no existing convention in which this expertise would be valued. To develop this skill would be like inventing a "private language" (though one could imagine an entrepreneur finding a way to have it taken up).

22. Goodman 1969 is relevant here but will be discussed at in chapter 5. Carlo Ginzburg's paper "Morelli, Freud and Sherlock Holmes: Clues and Scientific Method" (1989) treats connoisseurship as the ability to detect the author of a work of art and associates its method not with the "Galilean" sciences but with that of Sherlock Holmes, of medicine, and of historical scholarship in that all of them deal with specific instances of events rather than general relationships. Ginzburg seems confused. Physical and biological sciences, the identification of the provenance of paintings, and the method of Sherlock Holmes are all typically scientific activities: though they deal with specific instances, these instances are specific instances of general laws. In other words, general laws are applied to the specific cases under examination, just as, say, building a rocket to fly to the Moon applies general physical laws to a specific instance. History, on the other hand, is different, as Popper points out, since the broad flow of history is a one-off event; to believe that science can predict one-off events is what Popper (1957) calls "historicism."

how one must "negotiate" with the tiler over what will count as a satisfactory job in terms of both the formalized standards and the unformalized conventions. Interestingly, one may employ a professional—an architect—to do the negotiations. The fact that it is possible to employ a professional who may never actually have done any tiling to make these judgments shows that the crucial thing is experience within the conventions of judgment rather than experience of the skill itself. There is a connoisseurship of tiling. The judgment being exercised by architects, or homeowners, who themselves may not be capable of tiling (who have no contributory expertise) but who have seen and discussed many bathrooms, is based on *interactional expertise*.[23] Interactional expertise is the bridge between full-scale physical immersion in a form of life (which gives rise to contributory expertise) and non-expert acquaintanceship with the idea of tiling and the discourse pertaining to it. Interactional expertise enables architects to speak to both tilers and homeowners. The strong interactional hypothesis posits that by being immersed in the language community alone one may learn to "know what one is talking about" even if one cannot do the corresponding activity.

Is this, then, what Lord Campbell was getting at in his remarks about TV produces and sugar workers quoted in the introduction? No! Lord Campbell's view was different because he implied that he did not need interactional expertise or specialist experience of any kind (outside of management) to make the judgments. Just as the general public cannot have expertise in all domains of specialization, neither can any single person. There may be some who would claim that refined judgment in all things is the inheritance of the members of a well-bred aristocracy, but if the aristocracy does have special qualities in this regard it is actually a matter of training in the specific domains of food, wine, or art. When they take it to be their birthright to extend that refinement of taste to more technical domains things usually go wrong. The notion of connoisseurship does not, then, safeguard a Lord Campbell-type view, though it does make safe the idea that it is at least possible to judge an expertise without being able to practice it.[24]

23. We are grateful to Kevin Parry and Mike Bergelin for providing a lived example of tiling. As with the Wittgensteinian description of following a rule—it is not possible to completely describe following the rule, but it is possible to know when the rule has not been followed properly.

24. See Shapin and Schaffer 1987 for the class basis of what was counted as legitimate "witnessing" in early scientific experiments.

We can now see more clearly why it was that in order to be sure to rec-ognize that a solo violin-player was a fraud it would have to be a familiar piece of music that was being performed; it would have to be a piece of the general type in respect of which we were experienced listeners—the musical equivalent of bathroom tiling. Only in this way would the foun-dation of ubiquitous, or at least relatively widespread, experience on which judgment must be based be distributed among the population of non-musician listeners.

Downward Discrimination: Peer Review and Its Variants

The claimed superiority of peer review as a method of judging scientific papers, grant applications, and the like is based on the idea that the best judges of an expertise are those who share the expertise; in these areas it is considered that only those with contributory expertise should judge those with contributory expertise. But, and this may be little more than a truism, the medium of judgment, even when contributory expertise is used to judge contributory expertise, is interactional expertise. Quite simply, the reviewer of, say, a paper or grant application in gravitational wave physics is not exercising contributory expertise—he or she is not engaged in the physically involving act of detecting gravitational waves at the time the review is being written; rather, the reviewer is exercising interactional expertise—the ability to talk or write about gravitational wave physics.[25] Luckily, as we have seen, the relationship between con-tributory and interactional expertise is transitive: to have contributory is to have interactional expertise. If the interactional expertise is latent, it will have to be realized to the extent that the reviewer is going to make a useful comment on the paper or application.

Now, the possession of contributory expertise can be taken to guaran-tee that the maximum possible (latent) interactional expertise has been acquired, and that is a very good reason for taking contributory expertise to be a sound basis for judgment. A transitive relationship works one way; the possession of interactional expertise does not guarantee the posses-sion of contributory expertise, but, according to the strong interactional hypothesis, someone who possesses it in full ought to be as good a judge of the contributory expertise to which it pertains as someone who has

25. Of course, in one of the wider senses mentioned at the end of chapter 1, this is a contribution to gravitational wave physics.

the contributory expertise itself. In practice, however, the relationship is going to be more complicated. It is very hard in practice (though not in principle, as we have seen from Epstein's study) for someone with no contributory expertise to master the same level of interactional expertise as a fluent person with contributory expertise. So, in the main, but not always, those with contributory expertise will be (potentially) better judges than those without. The issue is confounded if the interactional expertise of those with contributory expertise remains largely latent—that is, if they lack interactional and reflective ability. In such a case a person with a great deal of interactional and reflective ability and a modicum of interactional expertise may turn out to be the better (though less than optimum) contributor to, say, a decision-making panel.[26]

A still stronger claim is sometimes made by art critics and the like. They say that a level of connoisseurship (which, as we have established, is itself based on interactional expertise) can be developed through assiduous viewing and discussing of art. They sometimes claim that this makes for judgments that are superior in principle to those of artists. For example, it might be argued that artists generally work in a narrow genre whereas critics have wide experience. Artists sometimes give implicit support to this kind of claim by refusing to exercise their interactional and reflective abilities, preferring to "let the art speak for itself."

In writing the above passage we have talked not just about the broad boundaries of potential expertise but about what might make one expert better than another. We have more or less said that, other things being equal, in the matter of judging an expertise "E," the more (realized) interactional expertise in E the better. This leads us onto dangerous ground, but it is ground that cannot be circumvented. It is dangerous because three decades of research in science and technology studies has shown us that *internal* judgments made by one expert about another are always contestable.[27]

Does this mean we have fallen into an epistemological trap? The answer is that if it is a trap it is a shallow trap. It must be possible to make certain internal judgments about expertise. If it were not, none of

26. Here, as we shall argue in chapter 5, lies an important difference between the sciences and at least some of the arts. In the arts, the locus of judgment favors realized interactional expertise and so it favors interactional and reflective ability more highly than do the sciences.

27. It is the very contestability of such judgments that make relational theories of expertise so attractive.

the comparisons we have discussed in the section on hoaxing and faking would make sense. In the absence of internal judgment it would make no sense to say that solo violin-playing is more difficult to fake than ability in avant-garde art because the difference between skillful and unskillful performances of all kinds would be impossible to notice. Life would be one long gamble with chance when it came to judging even a difficult expertise like violin-playing and clearly there is more to life than this. Confidence trickery and the like would cease to be a puzzle in need of explanation because it would come as no surprise that an unskilled person could pass themselves off as a skilled person—there would be no trick in it—there would be nothing to be explained. In other words, we could not make sense of the way we live our lives without some notion of internal assessment of expertises. So how do we make *internal* judgments about expertise that are more likely to be right than wrong?

The principle toward which we are working is what we will call "downward discrimination." We claim that judgments can be made within a discipline even though the judge's expertise within the discipline is very low; it can be done when those being judged have a recognizably lower level of expertise. Consider a claim made by a participant in the UK debate about the safety of genetically modified foods. The participant insisted that since radioactive tracers are used to mark genes during genetic manipulation then the genetically modified foods that result would themselves be radioactive. Nearly all readers of this book will understand that this claim is incorrect. They will be able to exercise downward discrimination even though, in many cases, it will be on the basis of almost no specialist knowledge of the particular science under discussion. It is just that the person making the claim has recognizably less knowledge.[28]

There is an important difference between external and internal judgment in this regard. External judgment does not have a preferred direction: it can be applied equally well upward, downward, or horizontally. That is to say, an ordinary person can reasonably distrust the demeanor or interests of even highly technically experienced and highly qualified spokespersons for the nuclear industry, or the tobacco industry, or any other industry, even if he or she is in no position to question their claims on technical grounds. With internal judgments the epistemological problem means secure judgment can run only downward, not upward nor

---

28. We are grateful to Matthew Harvey for this example, which is taken from his fieldwork observations of the UK debate.

horizontally. Where the direction is horizontal, there are only arguments and negotiations.[29]

Because it is impossible to make a technical assessment of the technical understanding of an expert with more expertise, those downwardly discriminated against may not recognize the validity of the judgment; higher levels of expertise may not be recognized for what they are. Well-founded downward technical discrimination is all too easy to confound with bias by those on the receiving end of negative judgments.[30]

Why doesn't the idea of downward discrimination simply return us to the old fashioned view of top-down scientific authority? Because it works only where there is a settled consensus. Thus, in a critique of the public understandings of GM foods we can exercise downward discrimination only in respect of those aspects of the argument where debate is long settled, not in respect of the technology of GM as a whole. For example, we can criticize such things as the claim that GM foods expose the consumer to radioactivity because radioactive isotopes are sometimes used as markers in GM laboratory experiments, but we cannot criticize the suggestion that insufficient testing has been done to guarantee that herbicide resistance will not spread to weeds and the like; in the latter case there is no settled scientific consensus to draw on.

To sum up, we all tend to believe we can make internal judgments of expertises upward, downward, and horizontally. The sociology of attribution is the study of the way actors negotiate the right to judge expertise; public legitimacy can be assigned to judgments made in any direction, and those judgments which do in fact gain public legitimacy gain it as an outcome of the interplay of power, alliance-building, and so forth. For example, in recent years the folk wisdom view has given a great deal of legitimacy to upward judgments while reducing the potency of downward judgments. The normative view that we are developing here is that internal technical judgments, which are of a good enough quality to contribute to science and technology policy, can be made only when they run downward.

---

29. This is the situation that holds between bathroom tiler and householder or architect. The architect is brought in not because he or she is better at recognizing good tiling but because his or her professional status can be invoked to settle what might otherwise be an endless argument about standards. The point is that the interactional expertise of householder or architect will not in itself settle the issue, it being applied horizontally at best, but it does give its possessor a place at the negotiating table.

30. All the judgments we describe may, of course, be wrong. That is the nature of judgment.

Referred Expertise

Another kind of reasonable internal judgment is based on referred expertise. Referred expertise is expertise taken from one field and indirectly applied to another. The term is taken from the idea of "referred pain"—for example when a back injury results in pain in the leg. Consider the managers and leaders of large scientific projects. In general they will not possess contributory expertise in respect of the many fields of science they must coordinate. Thus, Gary Sanders was first a professor in the field of high energy physics, then became the project manager of LIGO (the Laser Interferometer Gravitational-Wave Observatory), which turned on the very different science of interferometry, and at the time of writing has taken up the post of director of a major new telescope-building project—again a very different field. He remarked to Collins: "They give you the keys to the Thirty Meter Telescope on Day One and say, 'Drive it.' I found myself making key design decisions, not really knowing the history, the lore, the tradition, the lessons learned in the telescope." He added: "I'm not an observing astronomer. I have to listen to the arguments of 'planets versus galaxy formation versus stellar populations,' and 'this instrument should be a first light instrument and that instrument should be first,' and the campaign has already begun. . . . In the end, guess what?—The guy who's never spent a night on a mountain opening the shutter and doing an astronomy observation is going to say 'I selected [this approach] and these are my reasons.' What the hell is that?" Sanders explained the way he had learned during his first eighteen months on the project, using the vocabulary he had discussed with Collins:

> I was concerned that I just would not understand it. But I've found that, remarkably, what you call interactional expertise was not hard to achieve. I couldn't design an adaptive optics system but I really do, after six to nine months in the field, I really do understand the different kinds of adaptive optics and the way that they work and I can draw a schematic and define the algorithm, and understand the technological readiness of the different techniques—which ones are really ready to apply to the sky and which ones need to be demonstrated and certain components have to be developed. . . .
>
> I can sit down with a bunch of adaptive optics experts who will come to me and say "Gary you're wrong—multi-object adaptive optics will be ready at first light and it will give the following advantages . . ." and I shall say "No, it's multi-conjugative adaptive optics" and I can give them four

reasons why we should go with multi-conjugative adaptive optics based on the kind of science we want to do, the readiness of the technical components, when we need them, and so on, and I will see as I am talking about it that the room is looking back at me and saying "He does have a line, he's thought it through."

[But] if someone said to me, "OK Sanders, we agree with you, now go and design a multi-conjugative adaptive optics system," I couldn't do it. I couldn't sit down and write out the equations. . . . But I can draw a diagram of what each part does, where the technological readiness of each one is—what the hard parts are—I know the language and I actually feel qualified to make the decisions.

Looking back to his period at LIGO, he said:

I can't design the LIGO interferometer. I can't sit down and write down all the transfer functions and work out the noise budget like [named scientist] can. But if he gave a talk on it I could follow it. I can understand the important parts and the hard parts, partly by listening and partly by quantitatively understanding, but I couldn't come back and compose the symphony. But I was in a position where I had to decide. So it's a matter of who I listen to and which parts seem like they carry the argument—what it is that we want. . . . That's more than interactional but it's not quite contributory in, I think, your usual sense of the word.[31]

In most specialist domains in the field they have to manage, the managers, then, have interactional expertise but not contributory expertise.[32] Does this mean that their technical expertise is no greater than that of, say, a sociologist who has developed interactional expertise? To say "yes" seems wrong—as Sanders says, there is something going on that is a bit more than interactional expertise. The resolution seems to be that, although, as we can see, contributory expertise is not required to manage even the science of a scientific project, management does need kinds of expertise that are referred from other projects. The managers must know, from their work and experience in other sciences, what it is to have contributory expertise in a science; this puts them in a position to understand what is involved in making a contribution to the fields of the

31. Interview, 22 October 2005, Laguna Beach.
32. For examples of disagreement over whether managers from high energy physics had the competence to manage LIGO scientists, see Collins 2004a.

scientists they are leading at one remove, as it were. Managers of scientific projects with referred expertise would manage better (as well as with more authority and legitimacy) than those without it.[33]

The experience in other fields is applied in a number of ways. For example, in the other sciences they have worked in, they will have seen that what enthusiasts insist are incontrovertible technical arguments turn out to be controvertible; this means they know how much to discount technical arguments. They will know how often and why firm technical promises turn out not to be delivered. They will know the dangers of allowing the quest for perfection to drive out the good enough. They will have a sense of how long to allow an argument to go on and when to draw it to a close because nothing new will be learned by further delay. They will have a sense of when a technical decision is important and when it is not worth arguing about. They will have a sense of when a problem is merely a matter of better engineering and when it is fundamental. Interactional expertise is the medium through which this kind of expertise is made referable from one field to another.

We know that not all managers of scientific projects have referred expertise. General Groves, who ran the Manhattan Project, seems to be a case in point.[34] The question of whether you need referred expertise to manage a science is, presumably, related to the question of how much specialist knowledge you need to manage anything. If you believe that referred expertise is a good thing for managers, then to manage the making of "X" you need, at the very least, experience in making the closely related "Y." Does this work for Lord Campbell's position? Again, it probably does not because making sugar and making television programs are not closely related in this sense. Indeed, that is the very point of Lord Campbell's outburst.

Referred expertise, of course, is not the only kind of expertise needed by the manager of a scientific project. Such a manager also needs expertise in financial management, human resource management, networking skills, political skills, and so forth; some of these will comprise the contributory expertise of management itself. Crucially, a manager of a scientific or technological project will need local discrimination; they

33. Though in the case of LIGO, some scientists thought that the referral was from too distant a site. They thought that high energy physics, from where the managers came, gave them a misleading picture of the skills required to do interferometry: "What I found disappointing was that after two years the project manager still didn't really know what it meant to do interferometric detection of gravitational waves" (quoted in Collins 2004a).
34. See Thorpe and Shapin 2000.

will need to know how to judge, if not between the competing scientific arguments in the specialism, at least between the scientists in the specialism. The manager will have to listen to the competing claims of different specialists, each of whom will be more accomplished in terms of contributory expertise in the specialism, and judge between them as specialists as well as judging between their arguments.[35]

Meta-criteria: Criteria for Judging Expertises

Our goal, as explained at the outset of this discussion, is to find ways to separate those who fall into the envelope of potential judges in respect of various expertises from those who fall outside that envelope. Another way to try to do this is by reference to externally measurable criteria.

Credentials

The standard way to try to measure expertise externally is by reference to credentials such as certificates attesting to past achievement of proficiency. Possession of certificates will define a number of kinds of expert, but note that there are not credentials that indicate possession of many of the expertises we have discussed so far. There are no credentials for fluency in one's native language, nor for moral judgment, nor for political judgment. There are no credentials for ubiquitous discrimination, no credentials for the ability to distinguish between experts and novice violin-players, nor for the majority of other forms of connoisseurship (the exception being some kinds of professional roles that involve connoisseurship such as that of the architect). Above all, there are no credentials for experts such as the Cumbrian sheep farmers or the AIDS activists. Therefore we conclude that credentials are not a good criterion for setting a boundary around expertise.

---

35. Though bear in mind that, as sociology of scientific knowledge has shown, and as scientists acknowledge, judging the science even within an esoteric specialism often amounts to judging the scientists. For a sociological analysis see Collins 1992; for a scientist's remark, see Wolpert 1994, who says: "Scientists must make an assessment of the reliability of experiments. One of the reasons for going to meetings is to meet the scientists in one's field so that one can form an opinion of them and judge their work."

In the management of large scientific project, referred expertise can have advantages over contributory expertise; it carries less commitment to any particular way of doing things and can make for more unbiased decision-making (Collins 2004a). For further analysis of exactly what it is that the managers of big science projects do, see Collins and Sanders 2008 (forthcoming).

Track Record

Track record is a better criterion than credentials. The philosopher Alvin Goldman argues that track record of success in making sound judgments is a way for laypersons to choose between experts.[36] Reference to track record of success will certainly exclude a lot pseudo-experts but, again, it excludes too many. For example, it again excludes the sheep farmers and the like who might be applying their expertise to a technical debate in the public domain for the very first time. Likewise it excludes the ubiquitous and local discrimination of the public, for which no track records of success are available. Even when we get to qualified scientists and technologists, disputed expertise often concerns new fields for which there are no track records, fields in which track records take decades to establish, and fields in which the meaning of success is ambiguous. Track records, then, are only sometimes better than qualifications, and the "sometimes" are likely to be those where disputes are shallow rather than deep.

Experience

A criterion that does seem to set the boundary in a better place is experience in a domain. This nicely includes the sheep farmers, the AIDS activists, and all the other categories of expertise that we have described while excluding the general public from technical domains. We know from the outset that without experience within a technical domain, or experience at judging the products of a technical domain, there is no specialist expertise. Without experience of doing science, talking to scientists, playing or listening to violin-playing, or looking at and discussing bathroom tiling, the minimal standards for making judgments in these areas have not been met.

Thus, examination of the experience of, say, an Alan Sokal, would have been a better guide to the value of his work on "a transformative Hermeneutics of Quantum Gravity" than superficial peer review, and the same applies to Chauncey Gardiner, Hancock's avant-garde artist, and any number of confidence tricksters.[37] Confidence tricks and other such scams work when experience is attributed to the fraud on the basis of short acquaintance, whereas knowledge of their lack of experience would elimi-

---

36. Goldman 2001. See also Kusch 2002 on testimony.
37. See Sokal 1996.

nate them from the class of experts. (Though, of course, if they did have a lot of experience, that would not guarantee that they *were* competent.)

Periodic Table of Expertises Summarized Again

Let us now summarize what we have said about kinds of expertise. Once more we read down The Periodic Table of Expertises—Table 1.

Ubiquitous Expertises are acquired by all members of human societies in the course of the normal "enculturation" that takes place during upbringing. They include fluency in the natural language of the society and moral and political understanding. Ubiquitous expertises are the beginnings from which all other expertises are built.

Dispositions such as *interactive ability* and *reflective ability* convert latent interactional expertise into realized interactional expertise.

Specialist, or domain-specific, expertises include those with a relatively invisible component of ubiquitous tacit knowledge such as *beer-mat knowledge, popular understanding,* and *primary source knowledge,* and the full-blown specialist tacit-knowledge-laden expertise which enables those who embody it to contribute to the domain to which it pertains;[38] the latter is *contributory expertise.* The bridge between experts with contributory expertise and people who are not experts in the domain is *interactional expertise.* Interactional expertise is tacit-knowledge-laden expertise in the language of a domain, and it is acquired through enculturation in the domain language. Interactional expertise is the medium of discussion where technical judgments are made. In logic, there is a transitive relationship between the five specialist expertises, though it may not always be realized.

Meta-expertises are used for judging other expertises. *External* meta-expertises turn on the judging of skills through the judging of persons, or the more general characteristics of their discourse, rather than on domain-specific understanding. They include *ubiquitous discrimination* and *local discrimination* (which turns on local knowledge of people). *Internal* meta-expertise does depend on a degree of technical expertise within the domain. The most straightforward kind of internal meta-expertise depends on the application of contributory expertise to a domain through the mediation of interactional expertise. *Downward discrimination* applies even where a relatively low level of domain expertise is applied to a still

---

38. Remember, we include widespread tacit-knowledge-laden skills such as car-driving among the specialist expertises.

lower level. *Technical connoisseurship* turns on interactional expertise alone, which may have been specially refined for the purpose as in the case of certain kinds of professional or critic. *Referred expertise* depends on the indirect application of domain-specific contributory expertise from one domain within another.

Meta-criteria are attempts to provide externally visible indicators of expertise. We have argued that *experience* is the best of the three possibilities presented.

Problems of Categorization

The Borderline between Interactional and Contributory Expertise

We now have to deal with a conceptual problem. Interactional experts are continually making contributions to sciences in which we say they have no contributory expertise. Examples include the contributions of philosophers and sociologists, who have never written a computer program, to the science of artificial intelligence; the contributions of social scientists and statisticians, who have never examined a fingerprint, to fingerprint identification; the contributions of project managers, who have never designed or built an interferometer or a telescope, to the technology of interferometric gravitational wave detection and the design of large telescopes; and, more generally, the contribution of experts in science studies to scientific and technological debates in the public domain. Indeed, we will argue that there are cases when the potential contributions of those with interactional expertise but no contributory expertise are not sufficiently well recognized. So, when do the contributions of interactional experts turn them into contributory experts?[39]

A first step in the analysis is to distinguish between "making a contribution" and "being a contributory expert." Thus, those who drill the wells for the Shell Oil Company make a contribution to those who work in science studies by providing the fuel that gets them from home to office. Nevertheless, the well-drillers are not contributory experts in the field of science studies-they are contributory experts in oil production.

It may be that this is too easy a case because the contribution of the well-drillers does not touch on the core discipline of science studies. We can see this because their contribution would be the same irrespective of

---

39. This section arose out of the persistent queries of Simon Cole. Similar points are made in Selinger and Mix 2004.

the domain of the experts to whom the fuel was delivered; Shell's employees make similar contributions to nuclear power station engineering, ballet dancing, zoo keeping, and so forth. Nevertheless, the idea of "making a contribution" can also cover cases where there is a more direct link to the core domain but the relationship is marginal, or sporadic. Thus, based on his interactional expertise, Collins has occasionally made suggestions— rarely taken up but at least provoking discussion—about aspects of the science and technology of gravitational wave detection.[40] But a characteristic of interactional expertise is that it is parasitic on contributory expertise rather than freestanding, and Collins's rare contributions seem more like this than the kind of self-sustaining practical experience that could be passed on to others. As we have argued, even if there were a whole body of interactional experts like Collins (and nowadays there do seem to be whole communities of social scientists parasitical on the new genetics), their understanding and discourse would diverge from that of the practitioners as time passed unless it was continually maintained and refreshed by contact with the world of practice to which it refers.

Some of the other cases mentioned above might, however, have more of a contributory component. A discussion with Simon Cole about fingerprint identification revealed some of the possibilities. Simon Cole, an expert in science studies, has appeared as an expert witness in fingerprint cases where his evidence has been called to throw doubt on the certainty of identifications made by fingerprint examiners. In the cases in which Cole has been involved, his "book learning" has frequently been unfavorably compared to the practical experience of the fingerprint examiners. Cross-examinations of Cole exemplify the problem. For example:

Q: Your working knowledge of latent prints is actually minimal, isn't that right?

A: My knowledge is in how the profession developed and what's in their literature.

Q: I am going to ask the question again: Isn't it true that your working knowledge of latent prints is minimal?

A: If by that you mean by knowledge of how to examine latent prints and make comparisons the way that fingerprint examiners, do, yes it is minimal.[41]

40. Actually, on at least one occasion a suggestion was taken up.
41. People v. Hyatt, #8852/2000, Tr. Trans. 37 (Sup. Ct. N.Y. Kings Co. — Part 23 Oct. 4, 2001). The case is discussed at length in Lynch and Cole 2005.

In this case the court concluded: "What Dr. Cole has offered here is 'junk science.'"

In our language, for Cole, the problem represented by this passage of discourse is that the court recognizes only the practical expertise of fingerprint examiners as making a contribution to the domain of fingerprint identification. The argument is not merely a matter of legal expediency—the practical experts believe in their craft. Thus Cole reports that, at a conference, he asked a relatively friendly fingerprint examiner how he knew what he claimed to know. The reply was along the lines: "I wish you could come to my laboratory and learn to do what I do and see what I see, and then you would see why I know that I know what I claim to know."[42]

What we would like to bring about is the establishment of a discourse of expertise that would enable Cole, if he wished, to replace his defensive responses under cross-examination with a confident: "I do not have contributory expertise in the matter of fingerprint identification but I do have interactional expertise in that domain and this enables me to make a contribution." In due course we may imagine it becoming the ordinary occurrence for interactional experts to be allowed to speak alongside contributory experts.

Now, setting Cole's expertise aside for the moment, consider statisticians who believe the have something to say about the likelihood of a fingerprint match being correct. Like Cole, their warrant for claiming that they have something to offer can only have to do with their understanding of how fingerprinting works. If the statisticians did not know quite a lot about fingerprint identification, they would not be in a position to argue that their expertise was relevant to the procedure of the courts. Their warrant, then, turns on their interactional expertise in fingerprint identification practice. What the statisticians want to bring to court is, however, a self-sustaining contributory expertise: statistics can be taught in classrooms and transmitted from generation to generation in the absence of intimate contact with any realm of practice to which it might be applied. The case is like that of Wynne's sheep farmers, who possessed a relevant contributory expertise but insufficient interactional expertise to ensure that it was recognized, and that of Epstein's AIDS sufferers, who did develop the interactional expertise to make sure they were heard.

When SSK is applied to artificial intelligence, the same thing seems

---

42. These are paraphrases generated by Cole's recollection of the conversation.

to be happening. A contributory expertise which provides understanding of the social nature of knowledge is being brought to bear on AI, but it is interactional expertise in the practice of AI that puts the outsider in a position to argue that the social analysis of knowledge should be brought to bear. The interactional expertise, then, makes a contribution, the contribution being to establish the value of a novel contributory expertise in respect of the esoteric domain. The role of the interactional expertise is to argue for the value of the new contributory expertise.[43]

Such a process can lead to the novel contributory expertise coming to be a regular part of the esoteric domain. But nothing philosophically profound happens when such a transformation takes place; it is just that the boundaries of the domain have shifted. The statisticians, for example, still have only interactional expertise in the practices of the domain with respect to which they were interactional experts in the first place. They are in no more anomalous a position than, say, a contributory expert in the calculation of the strength of gravitational wave emissions from inspiraling binary neutron stars and a contributory expert in the quality of coatings on interferometer mirrors who have only interactional expertise in each others' narrow domains even though both can be described as contributory experts in the wider domain of gravitational wave detec-

43. In the early days of critiques of AI by outsiders, the AI community appears to have been highly resistant to "outside interference." Philosopher Hubert Dreyfus wrote the first and, arguably, still the most definitive book-length critique in 1972 (see also 1992). He reports that Marvin Minsky and Seymour Papert among others (central figures in AI) attempted to prevent him gaining tenure at MIT in consequence. The intervention of MIT's president was needed to rectify the situation (email to Collins, 5 December 2003). Subsequently, however, Dreyfus was invited to advise the US military on their AI projects, and the AI community has become in general much more open to outside criticism. For example, anthropologist Lucy Suchman, who wrote a very well known (1987) critique of AI while at Xerox PARC, subsequently remained there and built up a research group *within* the organization. In 2002 she was awarded the Benjamin Franklin Medal in Computing and Cognitive Science, sandwiched between awards of the medal in 2001 and 2003 to two of the most prominent figures in the community itself, Marvin Minsky and John McCarthy. Harry Collins was encouraged to develop his AI work (e.g., 1990, Collins and Kusch 1998), when on his very first attempt to critique the AI community (Collins, Green, and Draper 1985) he was given a share in the prize for best *technical* paper at the 1985 meeting at the British Computer Society Specialist Group in Expert Systems. Collins was subsequently invited to sit on an ESRC review committee distributing grants for research in AI and a Ministry of Defence panel. Of course, the heartland of AI, which is designing and/or building programs, easily absorbs less critically minded philosophers and was founded by psychologists among others. Also, in contrast to fingerprint identification, it has a tradition of internal critique (Weizenbaum 1976; Winograd and Flores 1986), perhaps because it is a university-based rather than a craft discipline (but it should be borne in mind that this did not help Dreyfus in the early days).

tion.[44] It is the interactional expertise of one narrow domain specialist in another domain specialist's expertise that makes it possible for the larger field of gravitational wave detection to exist—otherwise it would just be collection of isolated groups of specialists. To refer to a member of one of these narrow groups as a contributory expert in gravitational wave detection is a matter of choice of focus not of ontology of knowledge, because it is what you say you are an expert *in* that determines whether that expertise is interactional or contributory. What a contributory expert can be said to be a contributory expert *in* is, then, to some extent arbitrary, because what is counted as a "domain" is to some extent arbitrary. But this does not mean that an interactional expert in some narrow practical domain can become a contributory expert in that same domain just by changing the attribution.[45] The above argument can be represented in the cartoon (figure 4).

In the cartoon, circles with spikes are areas of specialist contributory expertise, such as mirror polishing, the calculation of wave forms, or fingerprint identification. The irregular lines are the boundaries of domains of expertise. The left hand domain is gravitational wave detection physics. It contains many specialisms linked to each other by their members' interactional expertise (the patches of conversation). A purely interactional expert like Collins is shown as a stick figure with the ability to produce conversational performance pertaining to the field which is indistinguishable from that of the others, but no part of Collins is found within any of the contributory expertise icons. The right hand solid boundary contains the fingerprint identification domain with, at the bottom, someone like a statistician who wants to become a regular contributor to it. The statistician is bodily immersed in statistical contributory expertise, and capable of linking this contributory expertise into the domain of fingerprint identification (solid boundary) via interactional

44. And bear in mind that there will be narrower practical specialisms even *within* the mirror-coating and source-strength calculation domains. The notion of "domain" is fractal-like (Collins and Kusch 1998, chap. 1).

45. The same analysis shows us that even though the managers of big science projects make obvious and major contributions to the outcome of the science, it is still possible to distinguish among their skills between the interactional, the referred, and the contributory. For example, in Collins and Sanders 2008 (forthcoming), the authors argue that Sanders's ability to know how much weight to give to a technical argument is a referred expertise, his ability to make good decisions in respect of adaptive optics is an interactional expertise, while his ability to draw up progress charts is a contributory expertise (in management). Once more, however, there is no reason to say that the combination of all these expertises is not simply full-blown expertise in the management of science.

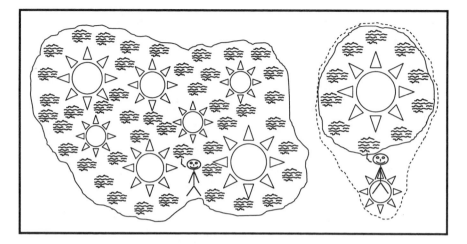

Figure 4. Contributory and interactional expertises in fields of science

expertise in fingerprinting. In the fullness of time the statistician's contributory expertise, along with the domain of statistics, will come to be seen as part of the field of fingerprinting (the ambition represented by the dotted boundary). The labels may change in such a case, and the statistician will become described as a contributory expert in fingerprinting, but his or her expertises will not have changed.

Exactly how a social scientist such as Cole should be represented on such a diagram depends on how his social science expertise makes its contribution. We can be sure that Cole can only justify any contribution he makes to the discussion of fingerprint evidence in court via his interactional expertise in the practice of fingerprinting, but whether his social science expertise might eventually become part of the domain after the fashion of the statisticians is less clear. Cole's expertise is more diffuse than that of the statisticians—it is a critique of the whole role of fingerprinting in court proceedings rather than a discussion of the correctness of any particular identification. Insofar as it has an impact on the court's decisions, it is the court that has to extract from his overall critique any contribution to the decision about guilt or innocence. Perhaps his kind of general critique is a one-off contribution which, once recognized, has no need to contribute actively to every case. In other words, while it may be that, although the critique arises from an expertise *sui generis*, it is something that is applied to different domains in turn, in the way that management consultancy is applied to different firms in turn. The social science critique is first applied to "DNA fingerprinting," then oral evidence, then old-fashioned fingerprinting, then other aspects of forensic science, and so forth. Neither the social science critique nor management consultancy become permanent "living" features of the domains to which they are applied.

SEE as a Normally Flawed Science

We have embarked on a categorization of expertise and, no doubt, many readers will already be finding even more faults with it than we have spotted. But all categorizations of expertise will be flawed—for example, there will always be boundary problems. One reason is that, as with any other categorization, it is necessary to deal with "ideal types." There will always be cases where one kind of expertise shades into another. Another reason is that experts often express their objections to a rival's conclusions by questioning their expertise. To give an example close to home, "science warriors" often say that sociologists' analyses of science are flawed because they do not have enough expertise in science.[46] These problems must be taken seriously but not to the point of academic paralysis. Social scientists should not aspire to a greater degree of perfection than the scientists they describe. Just as in natural science, many of the flaws in social scientific work have to be ignored if distance is to be allowed to work its enchantment—which it must if new knowledge is ever to be generated. This, of course, is not a way of avoiding assiduous critical scrutiny of our categories. The point is to understand the need for a table of this kind, either this one, a modified or elaborated version of this one, or one based on a new conceptual framework.

46. For typical work by "science warriors," see *Social Studies of Science* 29, no. 2 (1999), and Dawkins 1999; Gross and Levitt 1994; Gross, Levitt, and Lewis 1996; Koertge 2000; Wolpert 1992.

CHAPTER THREE

Interactional Expertise
and Embodiment

As explained, interactional expertise provides a bridge between the rest of us and full-blown physically engaged experts, and it touches on a wide range of professional activities. Indeed, the more one thinks about how our society works, the more one begins to see that interactional expertise "is everywhere." We now explore interactional expertise in more detail. The claim associated with the idea of interactional expertise is that mastery of an entire form of life is not necessary for the mastery of the language pertaining to the form of life. This is a big claim and needs strong proof. First we need to clear some ground.

The philosopher Hubert Dreyfus, among others, has argued that to learn a natural language you must be able to move around in the world, interacting with it physically, and that means you must have a body. The argument is central to Dreyfus's critique of artificial intelligence but goes against what we argue here.[1] Our question about interactional expertise can be expressed in the form: "How much of the language pertaining to a domain could a computer acquire in principle?" Our answer in respect of existing computers is almost the same as that of Dreyfus: we do not believe any existing computers have or could acquire human fluency in language. But Dreyfus's reasons are to do with the computer's lack of a body, ours are to do with its not having that part of the brain and the few other organs required for socialization. Dreyfus thinks that his position on the importance of the body rules out language acquisition by computers. The title of his influential 1967 article is "Why Computers Must

---

1. See Dreyfus 1972 and 1992 for his classic anti-AI arguments. The larger part of the anti-AI arguments put forward by Collins (e.g., 1990; Collins and Kusch 1998) are unaffected by what is argued here.

Have Bodies to be Intelligent." "Intelligence," as discussed in the article, includes natural language use. The position argued here is that you do not have to use your body (acquire contributory expertise) in order to speak the language of a domain (acquire interactional expertise), so Dreyfus's reliance on the embodiment thesis must fail as a critique of the possibility of computers acquiring language.

The key to the difference between Dreyfus's critique and ours, as mentioned in chapter 1, is that Dreyfus concentrates on the individual while we treat the location of expertise as the social group. That this is the location is most easily seen in the case of something like a natural language. The English language, for example, has "a life of its own," independent of the many native users of English which are the "substrate" on which it is instantiated. The only way to become a fluent English speaker is to be embedded in the English language-speaking community and absorb the ability. The only way to maintain the ability is to remain embedded as normal English usage changes. From this perspective—the perspective of this analysis—the crucial thing which prevents computers from becoming and remaining fluent English speakers is our complete lack of understanding of how to make them into regular members of English-speaking society. If we could make them into regular members of society, it would remove the crucial obstacle that prevents them from becoming fluent English speakers, French speakers, Zande speakers, gravitational wave physics speakers, or whatever, whether or not they have bodies.

We can go a long way towards showing this just by considering that humans can function well as natural language speakers under a variety of adverse circumstances in respect of their bodies. They can do this so long as their brains are still making sufficient connection with the embedding society to allow them to become and remain social beings. This, of course, does imply a minimal sensory apparatus, but the essential parts of the organism turn out to be very few. The essential parts are those bits of the brain to do with language-processing and those bits of the body to do with language-learning and speaking: the ears, larynx, and the rest of the vocal apparatus. In most cases, then, it is unsurprising that it is only the acquisition of *contributory* expertise, not interactional expertise, that turns on possession of a good proportion of the full set of body parts.

A further consequence of the new view is that those components of the body that are essential for embedding in a linguistic community, such as the larynx, are essential to the use of language in general, not the language of any specific domain such as English, French, or gravitational wave physics-speak. Again, this all fits with the separation of contributory expertise from interactional expertise.

Social and Minimal Embodiment

We now distinguish between the "social embodiment thesis" and the "minimal embodiment thesis." The *social embodiment thesis* holds that the particular language developed by any social group is related to the bodily form (or practices) of its members because bodily form affects the things they can do in the world. (It is a kind of inward-looking "Sapir-Whorf hypothesis.") This is what Wittgenstein seems to be getting at when he said that if a lion could speak we would not understand what it said. He was saying that we would not understand speaking lions because their physical make up is different to ours so the way they cut up the world conceptually—their "conceptual joints"—would be different too.[2] To exemplify, we understand the family resemblance between various things which we call "chairs" because we can sit on them because our knees bend in a certain way. Thus the word "chair" appears in our language and can be more or less mutually understood. A community of speaking lions, on the other hand, would not have the equivalent, or near equivalent, of "chair" in their language because they do not sit down in the same way. Instead, for lions, what we call a chair might be classed alongside whips and pointed sticks such as are used by "lion tamers" (assuming that the community of speaking lions still lived in circuses run by humans). Thus, here, a difference in the physical joints of the lions corresponds nicely to a difference in the conceptual joints.

Now we "roll out" interactional expertise. The idea of interactional expertise allows that a single lion with the capacity to speak, if snatched from its cradle and brought up alongside humans in the same way as are domestic dogs and cats, would acquire human language, including the word for chair, even though it, as an individual, could not sit. This is the *minimal embodiment thesis*. We call it the minimal embodiment thesis because it argues that though bodily form gives rise to the language of a community, only the minimal bodily requirements necessary to learn any language are necessary to learn the language of any community in which the organism is embedded. The minimal embodiment thesis follows from the idea that the language of a domain can be learned

---

2. Speaking lions would, of course, need larynxes and the brain development that goes with larynxes.

without full physical involvement in the domain. To repeat, interactional expertise allows that the language of a community whose members are embodied in one way can be acquired by individuals with bodies that are shaped differently and in ways that prevent them from participating fully in the physical activities of that community.[3]

If talking lions seem too bizarre, let us turn to another "thought experiment" invented by H. G. Wells. In his short story, "The Country of the Blind," Wells tells of a climber, Nunez, who tumbles down the side of a mountain into an otherwise inaccessible valley.[4] Nunez discovers that the eyes of all the inhabitants have atrophied, leaving nothing but sunken hollows. Thinking of the "well-known proverb," "In the Country of the Blind the One-eyed Man is King," Nunez expects to be respected and to become a leader. But, on the contrary, he finds himself treated as a clumsy fool. The inhabitants do not understand his talk of "seeing," of "stars" and the like, and take all this to be childish babbling or wickedness:

> For fourteen generations these people had been blind and cut off from all the seeing world; the names for all the things of sight had faded and changed; the story of the outer world was faded and changed to a child's story. . . . (474)

While Nunez tries to demonstrate his superior powers by making observations of movements from a long way away, the inhabitants of the valley are struck by his inability to apprehend what is happening behind walls and inside buildings, his senses of hearing and smell being nothing like as well developed as theirs nor his discrimination so fine. Likewise, they are appalled by his clumsiness in the dark, both at night and inside buildings which, of course, are not lit. They refuse to allow him to marry one of the women of the tribe in case he corrupts the race.

So far we see Wells exploring what we have called the social embodi-

---

3. The above section on talking lions and the different kinds of embodiment thesis draw on a couple of already published papers (Collins 1996a, 2000). Here the confusing term "individual embodiment thesis" has been abandoned, as the existing term "minimal embodiment thesis" serves the purpose better. Development of these ideas have been speeded by the critical comments of Evan Selinger (2003) directed at the earlier papers. The main ideas on interactional expertise have been published in Collins 2004b. Selinger and Mix (2004) responded, and Collins's reply, aside from what is here, is his 2004c.

4. Page numbers refer to the Odhams collected edition of Wells's works. This is undated and has no volume numbers, but the story runs from page 467 to 486 in the volume in which it appears. The piece was first published in 1911 in a collection of short stories, but publication details are unknown. The story is currently available in a Dover reprint.

ment thesis. In the country of the blind the whole way of life of the people has adapted to their blindness and so has their language. The ideas of seeing, of stars, and the like have atrophied along with the eyes of the people, while a new world of concepts has grown up which makes fewer distinctions between inside and outside and no distinctions between light and dark places.

Nunez, we are given to understand, once he realizes that in this place his sense of sight gives him more disadvantages than advantages, does try to moderate his claims. Wells lets us know that Nunez could, if he wished, adapt, with a struggle, to the ways of thinking and doing of the natives; to adapt or to remain unique becomes a matter of principle for the explorer, not a matter of ability. He falls in love with one of the women of the country and is torn between allowing his eyes to be removed—a condition set by the elders in hopes of curing his "illness"—and trying to escape. In the end he chooses a suicidal escape attempt, but at one time he had a choice. He could have chosen linguistic socialization; he could have managed this even without the operation that would make his body like that of the inhabitants of the valley. The attempt was made to force the blinding operation on him only because his ego continually trips him into proclaiming the advantages of sight in spite of its manifest disadvantages in this particular place—Nunez fails to "go native." The fact that he had a choice, and that he might have succeeded in acquiring the new conceptual structure were he not so obstinately determined to preserve his old way of thinking and acting, can be read as an illustration of the power of linguistic socialization. In spite of having a body unsuited to the place in which he found himself, Nunez could have acquired the language.[5] The counterpart of this story is the situation of the blind in our society, who seem to have little difficulty in acquiring the conceptual structure and native language of those who can see.

Madeleine

Turning from thought experiments to some real world observations, we come to the case of "Madeleine," described by Oliver Sacks in his book *The Man who Mistook his Wife for a Hat*.[6] Madeleine was born blind and

---

5. Exactly what Wells meant us to take from the story is not entirely clear. Was it perhaps a plea for the virtues of maintaining the colonialist outsider's view of the world and not to be corrupted by native values? It does not matter as far as we are concerned.

6. Sacks 1985, chap. 3.

disabled, being unable even to use her hands to read brail. Nevertheless, Madeleine learned of the world from books read to her by others. Madeleine had a minimal "body" with almost no ability to take part in the normal activities of the members of the surrounding society. Sacks's triumph with Madeleine was to teach her how to use her hands for the first time; the fact that it was a real *triumph* is established by his stressing the extent to which Madeleine had been utterly *inactive* throughout her earlier life. The uselessness of Madeleine's hands, according to Sacks, had come about precisely because she never did any moving about in the world on her own. Sacks says "she had never fed herself, used the toilet by herself, or reached out to help herself, always leaving it to others to help her" (58). Nevertheless, Madeleine learned to be a person who "spoke freely indeed eloquently . . . revealing herself to be a high-spirited woman of exceptional intelligence and literacy" (56) through the medium of the written and spoken word. Madeleine, according to Sacks, was "of exceptional intelligence and literacy, with an imagination filled and sustained, so to speak, by the images of others, images conveyed by language, the *word* . . ." (59, Sacks's stress). Madeleine, it seems, is exemplifying the minimal embodiment thesis and the power of interactional expertise. She has learned the language through immersion in the world of language alone rather than immersion in the full-blown activity which constitutes the form of life. The social embodiment thesis says, correctly, that the language has arisen from that full-blown form of life—that is, from the full range of activities of the full-bodied members of the society— but the case of Madeleine shows that an individual can get much of the corresponding understanding without much of a body.

If a minimal body only is required to acquire a full-blown language pertaining to a full-blown form of life, how minimal can this body be? This question seems unlikely to be asked by philosophers such as Dreyfus who treat embodiment as, as it were, a binary quality. One either has a body or does not have a body—the values are "1" or "0." If this approach is taken Madeleine's symbol would be "1" rather than "0." Our view is that the binary approach leaves unexplored the huge terrain of interactional expertise. To open up the territory it is necessary to ask questions about how much physical involvement is needed for linguistic involvement, and we can go a long way toward answering this question by treating the body as continuous—something of which there can be more or less—and examining how much is needed in order to develop linguistic fluency in various domains. This is the kind of question to which we can bring empirical techniques rather than *a priori* philosophical argument. We prefer to see the case of Madeleine as posing an empirical question and, if Sacks's description is exact, providing an empirical answer.[7]

The Prelingually Deaf

Though the body can be minimal, the theory of interactional expertise does not allow it to be absent if language is to be acquired. What is necessary is the equipment for engaging with the linguistic community of a host society. Fluency in language cannot be acquired unless there are the physical means of engaging in the linguistic to-and-fro that is linguistic socialization as opposed to being fed with propositions in the form of something like a computer program. The point is illustrated by the case of the congenitally or prelingually deaf.

The prelingually deaf—those who are born profoundly deaf or who become profoundly deaf soon after birth and before they have acquired the beginnings of language—have perfect mobility, and every other faculty, but lack the ordinary means of linguistic socialization—speech. In case it is not immediately obvious, the normal medium of language-learning is the spoken word—that is how babies learn; reading and writing come very much later and are parasitic on a language already learned through speech. It is possible to learn a form of reading and writing without speaking. For example, young orthodox Jews are taught to read ancient Hebrew prayers phonetically. But, as at least one of the authors can attest, this is a long, slow, and painful business, which is in no way to be associated with learning a *language*. It is, rather, like learning to chant something meaningless with the aid of written cues.[8]

It is not impossible for the prelingually deaf to learn ordinary language, but it can be done only with an immense effort and with the aid of teachers who will spend hours a day, using lip movement and other non-aural cues, to substitute for the bath of speech in which the infant is normally immersed.[9] Unfortunately, the prelingually deaf child usually

---

7. For more discussion of the relationship between philosophical and empirical approaches to this question, see Selinger, Dreyfus, and Collins 2008.

8. Something similar must also apply to the Catholic church where, or when, prayers were conducted in Latin. One might also liken the situation to Searle's notion of the "Chinese Room" (the Hebrew Room?).

9. Sacks describes one of his colleagues who was so fluent at lip reading that he did not realize she was congenitally deaf. But her mother had "devoted hours every day to an

misses out on early language socialization and as a result does not learn to master the natural language with any degree of fluency.

Mistaken views of the situation of the congenitally deaf were embedded in an educational program known as "oralism." This held that the prelingually deaf could be taught ordinary English via written signs and lipreading; it gave rise to an education program in which signing was banned. Ladd recounts the experiences of a young boy taught under this regime:

> They'd write on the board, and we would copy it. Then they would give you good marks and you would swagger about. But what did those words mean? Ha! Nothing! It all went past us. . . . Yet those two-faced people would give us good marks and pat us on the head. (Ladd 2003, 305)[10]

The outcome of oralism, according to Ladd, was that the prelingually deaf had an average reading age of about eight when they left schools, with consequent failure to attain significant positions in society or make their way in the world.

Ladd draws a distinction between "Deafness," where the capital "D" signifies "Deaf consciousness" along with congenital deafness, and other forms of disablement. Those who are merely "hard of hearing" or "deafened" (later in life) do not share a common consciousness with the (large-D) Deaf because they do not have the same language problems:

> It is vital to note that ["small-d" deaf] people share English as a first language with disabled people, and thus there are few communication or cultural barriers. Because these are the people with which the disability movement comes into contact, it is easy for them to mistake the reality of Deaf communities, who face the same linguistic/communication barrier in interaction with disabled people as they do with anyone else. (168)

The large majority of Deaf people, then, are not good at reading written English though they can see it perfectly. Ladd writes that "they have little access to written English discourse" (55) and that "most organisations still print their information in English, even whilst knowing that most Deaf people cannot read it" (179).

---

intensive one-to-one tuition of speech—a grueling business that lasted twelve years" (Sacks 1989, 2 n. 4). This possibility, of course, allows the minimal body required to master a language to become even more minimal.

10. The experience of the deaf child taught English in this way resonates with the teaching of Hebrew just described.

To summarize, there is no reason to suppose that Sign is deficient as a language. In some respects, such as the way it heightens spatial acuity, it is superior to spoken languages such as English. Nevertheless, it is different, and this has important consequences for members of the Deaf community. One of the most significant is that, although surrounded by written texts—the visual counterpart of the spoken language, as we might say—members of the Deaf community find reading and writing difficult. The reason is that, for the Deaf, these texts are not grounded in the native spoken language they represent and, as such, solve no problems. Not being able to hear the spoken language seems to be a greater obstacle to learning it than, as in the case of the blind, not being able to see its visual representations.[11]

We can also see that insofar as a body is required to participate in a linguistic community, then it must include some physical structure that allows it to open itself to the social world of that community. In most cases this will mean the ability to hear and make sounds, but we know from the experiences of the Deaf that even the loss of the part of the body that normally has the responsibility for language acquisition can be circumvented and alternative linguistic communities, such as those involving sign, can grow. It is also the case that, if enough is done with the senses of sight and touch, then, like the explorer in H. G. Wells's story, the prelingually deaf individual can grasp the conceptual structure of the native "oral" society if they are given enough special attention from an early enough age. In the normal way, however, this does not happen, and the response of the Deaf community has been to stress their difference as a community.[12]

Interactional Expertise and Practical Accomplishment

Taken together, these examples show the importance of linguistic socialization relative to full-scale physical immersion in a culture. Let us try to make what we have argued still clearer. We are trying to establish what would happen in a kind of neo-Turing test or "imitation game." In his

11. Those with multiple disabilities, such as Helen Keller, report that loss of hearing is far more devastating than loss of sight in the way it cuts them off from the world.

12. Ladd argues that for the Deaf to fit in with society at large it will be necessary for hearing persons to become bilingual in Sign (as they are in isolated locations such as Martha's Vineyard), or (as Ladd suggested in a personal communication), for more individual attention to be paid to the training of the Deaf from an early age so that they could become bilingual.

famous 1950 paper, Alan Turing proposed that a hidden computer compete with a hidden human; a "judge" would try to determine which was which in the course of an interrogation conducted through keyboards. The idea was based on a parlor game—the "imitation game"—in which a hidden man tries to pass himself off as a woman while answering written questions, the answers being compared with those of a real woman.[13] Imagine that Madeleine were placed behind one curtain and a fully mobile person were placed behind another. Let us then imagine that a fully mobile interrogator were to engage each in written conversation. The claim made in this chapter is that the interrogator would have a hard time distinguishing who was who.

As was intimated in the introduction, we have three ways of learning to succeed in an imitation game where the medium is ordinary conversation. There is (a) full physical immersion in a form of life (which, of course, eliminates the need for "imitation"); there is (b) linguistic socialization in the absence of shared physical activity; and there is (c) feeding with discrete propositional knowledge. We believe that method "c" will fail and agree with the philosophical arguments that lead to this conclusion. We believe, however, that there are questions still to be asked about method "b." Methods "a" and "b" have not been distinguished in the existing philosophical/sociological discourse. What we are claiming here is that that method "b" can accomplish as much, or nearly as much, as method "a," where the "achievement" is measured by conversational ability (and therefore ability to contribute to judgments within the form of life).

Note something important that is not being claimed here. It is not being claimed that mastering the language of domain through method "b" provides the practical capacities belonging to that domain that would be mastered alongside the language if the method of language acquisition was method "a." Learning the language of a domain is not a substitute for learning a whole form of life. That is why we stress that learning a language via method "b" is indistinguishable from learning it via method "a" only if the test for indistinguishability is carried out via the medium of language—as in an imitation game. The point is, however, that much in the way of human affairs is carried out through the medium of language alone so that a language mastered through method "b" can be very useful and powerful. That is why interactional expertise is so important and widespread.

We have suggested that the fact that a language can be learned by

13. Turing 1950; Collins 1990, chaps. 13, 14.

method "b" has not previously been noticed because, in the philosophical literature, method "b" is elided with either method "c" or method "a."[14] The first elision takes it that linguistic socialization is the same as feeding with propositions and therefore linguistic socialization is taken (implicitly or explicitly) to fail. But linguistic socialization is very far from feeding with discrete propositions; the language learned as a result of linguistic socialization is as loaded with tacit knowledge, Wittgensteinian rules, and ability to make intuitive judgments, as any native language.

The second elision is that made by Dreyfus.[15] It is that the reason that method "a" and method "b" are not distinguishable in a Madeleine-like case is because Madeleine, in spite of her immobility, is to be counted as substantially physically immersed in the form of life even though she uses almost none of her body in the course of her life. Dreyfus says that even though Madeleine does not have the use of her legs, arms, eyes, and so forth, she still has a front and a back and can be moved about in the world. This is "the body as binary" argument and, as mentioned above, it is immune to empirical falsification and leads in the wrong direction to allow practical consequences to flow from it. Madeleine's bodily state ensures that she has no contributory expertise in any practical activity, so any acquisition by her of normal-looking discourse proves the point we are trying to make—interactional expertise can be learned in the absence of contributory expertise.[16]

14. We can apply the idea to another standard philosophical thought experiment: Monochrome Mary (see Jackson 1986 and http://www.calstatela.edu/faculty/nthomas/marytxt.htm). Monochrome Mary lives in a entirely black and white world. Philosophers are interested in whether Mary, if she could master colorful discourse, could be said to have knowledge of color. In the philosophers' treatment, Mary becomes a perfect color scientist by garnering all the propositional knowledge about color there is; she comes to know everything there is to know about the physics, physiology, and psychology of color. Our question is slightly different, however. We are asking under what circumstances Mary could pass in a Turing Test-like situation when interrogated by a color perceiver. We are suggesting that if Mary were nothing more than a perfect color-scientist—where "perfect color scientist" is taken to mean knowing all there is about color that can be written down—then Mary would fail. If, on the other hand, Monochrome Mary had been brought up holding conversations with color-perceiving speakers, then she would be hard or impossible to distinguish from them in Turing-test like circumstances. (We are grateful to Evan Selinger for drawing the relevance of Monochrome Mary to our attention.)

15. See also Selinger 2003.

16. In his description of Madeleine, Sacks claims to have improved her ability to engage physically with the world and, for example, to use her hands to feel things and especially to model clay sculptures of people's heads. While not wishing to ignore this achievement, what is of significance to us is that Madeleine was able to speak so fluently about so much before this transformation took place.

To repeat, our argument is that the sociologist is a kind of Madeleine in respect of, say, a group of gravitational wave scientists, and has only linguistic socialization to draw on. Nevertheless, the sociologist can still perform reasonably when tested in conversation about gravitational waves, just as Madeleine can perform reasonably well when tested on her ability to maintain nonspecialist discourse.[17] Our argument, of course, is not just about sociologists but about all those professionals who fill, as we might now see it, a Madeleine-like role.

Going back to the other examples referred to in this chapter, the blind, and other disabled groups, work toward embedding themselves in the surrounding culture without difficulty because they are thoroughly linguistically socialized; the Deaf, in consequence of the difficulty for them of linguistic socialization, prefer to value their identity as a separate community with their own language.[18] The social analyst, to repeat, appears less like the Deaf and more like the wheelchair-bound, or the blind in our society, or the explorer in H. G. Wells's story—capable of mastering the unfamiliar conceptual structure of the new surrounding society given enough linguistic immersion, even without physically engaging with the practice of the target society. The minimal embodiment thesis is the model which should inform our activities and understanding, not the social embodiment thesis. The talking lion snatched from its cradle and brought up in human society, Nunez the explorer in the Country of the Blind, Madeleine, and the sociologist of gravitational wave physics should all be able to succeed in imitation games where the judges are fully socialized native language speakers, whether in ordinary human language (the lion and Madeleine), Country of the Blind language speakers, or gravitational wave physics speakers. On the other hand, if we are thinking straight, natives of the Country of the Blind, if interrogated by a seeing person, should fail because they have had no opportunity to master the seeing person's conceptual world. Remove the linguistic socialization, as in the case of the Deaf, and the potential for conceptual mastery is severely prejudiced irrespective of the degree of embedding in a native

17. We have to bear in mind throughout this chapter than insofar as we are using Madeleine as our exemplary case, all we know about the actual Madeleine is contained in Oliver Sacks's short and popular account. If it turns out that Madeleine was really a much less fluent speaker than Sacks implies, then the quasi-empirical support for the argument made here would be weak. Still, the first thing is to shift the agenda from the purely philosophical to the empirically testable. In the next chapter we will discuss experiment proper.

18. This is not to say that groups of the blind and the wheelchair-bound may not choose to develop political and cultural consciousness, but the imperatives would seem to be different.

community. We will describe tests of these claims below. To finish our philosophical critique, however, let us return to Dreyfus.

A Note on the Body and Lenat

The discussion of Madeleine arises because it was central to an earlier debate between Hubert Dreyfus and computer enthusiast Douglas Lenat. It was Lenat who introduced Madeleine in opposition to Dreyfus's claims about the need for a body if a computer was to be intelligent. He argued that if Madeleine did not need a body to learn human language, neither did a computer. In this argument we find ourselves agreeing with Lenat at least to this extent: if it is necessary for a computer to have a body in order to know language, it is not much of a body. The crucial features of the body are to do with communication with the rest of the language-speaking community. These features are the vocal apparatus, the ears and the parts of the brain associated with them. The front and back, the inside and the outside, the ability to be moved around may or may not be necessary. But if they are necessary, it is because, like the vocal apparatus and so forth, they are needed for language acquisition in general, not the acquisition of any specific language, such as the language of the blind or the language of gravitational wave physics. In sum, the arguments of Dreyfus and his supporters about the importance for language acquisition of those basic elements of the body which are not to do with speech show nothing about the importance of the body for the acquisition of specialist languages. Consider the examples that Dreyfus took from Merleau-Ponty in his 1967 article —the blind person's relationship to the stick, and the matter of feeling silk. Front, back, and movement are not essential to the acquisition of "blind person's stick language" or "feeling silk language" in particular. Therefore the idea that only a person who has the bodily apparatus to engage in the practical activities of a specialist group can acquire that group's language has not been proved by Dreyfus and his supporters.

None of this is to say that Lenat's project, CYC, which aims to enable a computer to accomplish normal interaction by filling it with all the knowledge that could be found in encyclopedias, is anything other than mistaken in its premises. Clearly, a computer filled with every bit of propositional knowledge that could be expressed has not been linguistically socialized: it has none of the tacit components of language that enable linguistic creativity, repair of damaged speech, and so forth. It is simply not an element in the larger social organism that is the substrate of language.

This is not to say that Lenat's recent ambition, of locating CYC in the Web where it can learn masses of information from other Web-users, is a hopeless idea; it is just that it will be accumulating information, not language-speaking abilities. We would claim that it would still fail a Turing test in which true linguistic socialization, such as repair of broken language, was at stake. The same analysis applies to all imagined devices where language is embedded as a once-and-for-all capacity, such as Searle's Chinese Room and the device imagined by Ned Block.[19] In each of these cases the crucial components are missing—the components that link the device into the larger body of society.

To sum up the philosophical critique, the implication of the idea of interactional expertise is that we need to examine the requirements for the body more closely, because it looks as though they are minimal. On the other hand, we need to take much more seriously the requirements needed to acquire a "social fluency," such as language, which we may think about as being more to do with the purely language processing parts of the organism than had previously been argued. We can think further about these questions by asking what any entity learning language through socialization can do without. We know it is possible to take away a lot of the body and still leave the entity with good language-understanding abilities—exactly how much we can take away is the question posed by what we might call "the minimal embodiment hypothesis." The importance of the minimal embodiment hypothesis for our project is that it supports the strong interactional hypothesis—that fluency in the language of a domain can be acquired outside of bodily engagement with the practices of the domain: if only a minimal body is required to acquire language, then this must be so.

19. See the "Editing Test" discussed in the coda to chapter 4 below. For a discussion of Block and the like, see Collins 1990, chaps. 13 and 14.

CHAPTER FOUR

Walking the Talk:
Experiments on Color Blindness,
Perfect Pitch, and Gravitational Waves

The strong interactional hypothesis implies that complete socialization is indistinguishable from thoroughgoing linguistic socialization when they are compared through the medium of language. Unfortunately we don't have ready access to talking lions, to a Country of the Blind, or to a series of Madeleines. But we do have access to the equivalent of the Country of the Blind or, to be exact, to two equivalent countries. What we have done is test the strong interactional hypothesis with some purpose-designed experiments, initially in cases where it really ought to work if it has any validity. We investigate "The Country of the Color Perceivers," into which have fallen a few intrepid explorers—those who are color-blind. "The Country of the Color Perceivers" is, of course, our own country. Our own country can also be described as "The Country of the Pitch Blind" since hardly any of us can recognize absolute musical pitch. Just a very few possess the facility of "perfect pitch." In the case of pitch it is those who can "see" pitch, as opposed to those who cannot see color, who take the role of Nunez.

About five percent of males have no contributory expertise in red-green discrimination. Nevertheless, they *have* been immersed since birth in the language of color, so they should have maximal interactional expertise in color language even though they do not have contributory expertise. The strong interactional hypothesis holds that color-blind persons should succeed in pretending to be color-perceivers in the imitation game. It is supported by the fact that those around them generally notice no deficiency in color-blind persons' color-talk; color blindness often goes undetected in the absence of purpose-designed tests of contributory expertise.[1]

---

1. Selinger argues that red-green color-blind persons can perceive other color differences, and thus already possess some contributory expertise in color perception. By

"Just as you see an apple and know it's red without thinking about it, I hear a note and know it's an E flat." That is a description of perfect pitch. In contrast to color blindness the "disability" is the statistical norm. Therefore "pitch-blind" persons (nearly all of us) have not been socialized into the language of pitch-perception. The strong interactional hypothesis holds, then, that it should be easier to spot the pitch-blind pretending to be pitch-perceivers in the imitation game than the color-blind pretending to be color-perceivers.

The same logic suggests that "normal" people should be bad at imitating the color-blind because they have not been immersed in the linguistic world of the color-blind (as in the case of natives of the Country of the Blind trying to imitate seeing persons). Here we treat the ability of a color-perceiver to pass as a color-blind person as the expertise to be imitated. Only the color-blind will know "the tricks of the color blindness trade" and the discourse that goes with it. Using the same logic, those with perfect pitch should be good at imitating pitch blindness because they have grown up surrounded by pitch-blind persons.

We can now report on initial experiments intended, among other things, to make the idea of interactional expertise more concrete.[2] The idea on which the experiments are based, the "imitation game," is more than fifty years old, being the forerunner of the famous Turing test. We use the imitation game to investigate the discourse of people who do not possess a certain skill but have been immersed in the language. To do

---

extension, this is an argument against any experiment of this kind, since any contributory expertise at all could be said to bear on any interactional expertise. It returns to the philosophy of the body argued by Dreyfus in his debate with Lenat—namely even the most minimal body is, nevertheless, a body: embodiment is, as it were, a binary quality, and Madeleine's symbol was "1" not "0." As argued above, we think it is more useful to think of the notion of a body as continuous and we believe we are doing experiments which do bear upon the credibility of the Madeleine case as described by Sacks as well as on the notion of interactional expertise. The idea of interactional expertise dissolves if embodiment is a binary quality, since all discourse among humans is embodied and, logically, experiments of our kind, irrespective of the four-way control conditions and clear outcomes, can reveal nothing of significance. For the determined philosopher, then, the question becomes: what do these experiment reveal and what would change had the results been different? (See Selinger, Dreyfus, and Collins 2008.)

2. Experiments can have a function that goes beyond their findings. Harvey (1981) shows that initial experiments on the problem of non-locality in quantum mechanics did much to develop the ideas even though the results were not treated as decisive. Collins (e.g., 2004b) argues that Weber's failed observations nevertheless settled a long-running theoretical argument about whether gravitational waves could be detected in principle and went on to found the half-billion dollar field of gravitational wave detection.

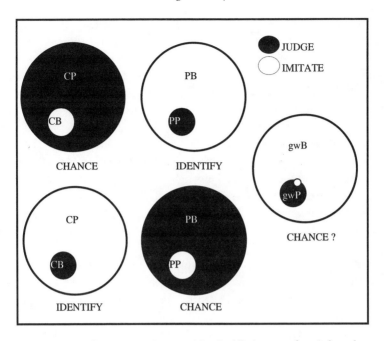

Figure 5. Socialization experiments with color blindness, perfect pitch, and gravitational waves

this we use the imitation game to compare the domain-specific linguistic abilities of interactional experts with that of contributory experts and that of nonexperts. The experiments are represented in figure 5 above. The four circles on the left represent tests on color blindness and perfect pitch which are used to explore the concept of the experiment. "Proof of concept" having been accomplished, the rightmost circle represents a first use of the idea in an area of more direct relevance to scientific and technological expertise, namely as an indicative test of the ability of a participant observer to pass as a scientist.

The first circle in the top row of figure 5 represents a society like ours in which the majority of people are color-perceiving (CP); that is, they are not color-blind. A minority are color-blind (CB). The idea of interactional expertise implies that, having been brought up in color-perceiving society, the color-blind will be fluent in color-perception language even though they cannot see the full range of colors—they will have acquired interactional expertise in color-perception language though they have no contributory expertise in color discrimination. Therefore if the discourse of a color-blind person, who is trying to imitate a color perceiver, is compared to the discourse of a color perceiver, even if the judge is a color perceiver, no difference should be detectable and the judge can do no more than guess. The success rate of these guesses should be no better than chance.

The second circle represents a society, also like ours, in which a small minority are "pitch-perceiving" (PP), that is, they have perfect pitch. The majority are "pitch-blind" (PB). The theory holds that in this case the pitch-blind will not have mastered pitch-perception language because they have not been brought up among pitch perceivers and have not been immersed in pitch-perception language. In other words, we should not expect the pitch-blind to have acquired interactional expertise in pitch-perception language. Therefore, if the language of a pitch-blind person who is trying to imitate a pitch perceiver is compared with the language of a pitch perceiver, and the judge is a pitch perceiver, the difference should be detectable. We ran imitation game experiments to test these hypotheses. As can be seen, in figure 5, white signifies the group from which the imitator is drawn while the dark always represents the groups from which the judges are drawn.

The bottom two circles in the figure are like the top two circles except that the "polarity" is reversed and the results should be the opposite. In the first circle of the bottom row color perceivers try to imitate the color-blind, and we should expect them to fail so that a color-blind judge can identify who is who. In the second circle pitch perceivers try to imitate the pitch-blind, and we would expect them to succeed so that a pitch-blind judge (most of us) cannot identify them and the judge's guesses should, once more, be random. Since we know what we should expect in these cases if the idea of interactional expertise makes sense, the experiments on color blindness and perfect pitch can be treated as "proof of concept."

The rightmost circle in figure 5 represents an application of the concept and the experiment to science studies and, in effect, participatory, anthropological, and ethnographic fieldwork in general. Here a member of the wider society enters an esoteric group in an attempt to acquire its interactional expertise. In this case the esoteric expertise belongs to gravitational wave physics. The large majority of members of our society are, as it were, "gravitational wave physics–blind" (gwB)—they have no deep knowledge of gravitational wave physics. A small minority are, as it were, "gravitational wave physics perceivers" (gwP). The small white circle represents a member of gwB society who enters the black circle of the gwP, hoping to learn the language without learning the practice of the physics. The final set of experiments described test whether the person represented by the small circle has succeeded in acquiring the targeted interactional expertise. The person tries to imitate the language of gw physics and the test is whether judges who are gw physicists can tell the difference between the participant's answers and those of gw physicists who do have contributory expertise. Success would be indicated by a chance outcome, failure by ready identification.

Procedure and Results: The Proof of Concept Experiments

Alan Turing's famous definition of intelligence in a computer (1950) turned on what has become known as the Turing test: a hidden computer and a hidden person would be interrogated by a judge via teletypes. If after five minutes or so of interchange the judge failed to identify the computer, it would be deemed to be intelligent. As explained, the particular imitation game on which Turing's test was based was a parlor game in which a judge asked written questions of a hidden man pretending to be a woman and compared these with the answers of a hidden woman who replied honestly. In our terms, if the hidden man were to succeed in fooling the judge, he would have demonstrated the possession of the interactional expertise associated with being a woman though not the contributory expertise.

This is the protocol we applied to color blindness and perfect pitch, in this case using purpose-built software to link three computers via a wireless network.[3] Judges sit at one computer and can type any question that they think will probe for possession of the target expertise (let us say color perception as in the first circle of figure 5). The question is transmitted simultaneously to both participants, one of whom will be color-blind but pretends to be a color perceiver and one of whom will be a color perceiver who is instructed to answer "naturally." When both participants have replied, the answers appear simultaneously and side-by-side on the judge's screen. The judge can then make a guess and provide a "confidence level" associated with the guess. The judge is then free to ask another question. The session continues until the judge feels there is nothing further to be learned by going on. In our experiments judges usually felt that there was nothing more to be learned after they had

3. More details of the method will eventually be posted on the Web in the form an experimenter's handbook. See Collins 1990, chap. 13 for an explanation of the rationale behind the judges possessing the target expertise.

asked around half-a-dozen questions, though some asked fewer and one or two asked many more.

Judges

It seems obvious that judges will have the best chance of identifying the participants if they possess the "target expertise"—the expertise that is the subject of the pretence. For example, in the gender imitation game, if it was the hidden man pretending to be a hidden woman then the judge should be a woman. A male judge's model of a woman, we suggest, is likely to be similar to a male "pretender's" model of a woman, and this would make it too easy for the man to pass. Thus, in our tests, if the participants are pretending to be color perceivers the judge will be a color perceiver; if the participants are pretending to be color-blind, the judge will be color-blind.

Though the need for judges to possess the target expertise seems obvious, it is worth dwelling on. It bears on the significance of "knowing what you are talking about." If judges who did not know what they were talking about could easily tell who was an expert and who was not, purely by examining the technical contents of discourse, it would suggest that upward discrimination was a possibility. It would suggest that the value of knowing what you are talking about may have been exaggerated. For this reason we tested the importance of judges' possession of the target expertise—what we call Phase 3 of the experiment—and report the results of the test below.

Chance and Identify Conditions and Outcome

We refer to runs in which we do not expect the judge to be able to identify the participants as the *chance condition* and the runs in which we do expect the judge to be able to work out who is who as the *identify condition*. Table 3 puts into words what is represented in the left hand circles of figure 5.

Table 3: Expected outcomes of socialization experiments

	Pretender is		Target expertise	Expected outcome
A	Color-blind	imitates	Color-perceiving	Chance
B	Color-perceiver	imitates	Color-blind	Identify
C	Pitch-perceiver	imitates	Pitch-blind	Chance
D	Pitch-blind	imitates	Pitch-perceiving	Identify

We have four experimental configurations setting up chance and identify conditions for each of the color blindness and perfect pitch experiments.

The strong interactional hypothesis is demonstrated if guesses are no better than chance in the chance condition while this is not so in the identify condition—which can be thought of as the "control group." The idea of interactional expertise is demonstrated whenever the proportion of right guesses is greater in the identify conditions than in the corresponding chance conditions.

Confidence Levels

Judges were asked to use a four-level scale to record their confidence after each guess:

- *1st level:* "I have little or no idea who is who."
- *2nd level:* "I have some idea who is who—but I am more unsure than sure."
- *3rd level:* "I have a good idea who is who—and I am more sure than unsure."
- *4th level:* "I am pretty sure I know who is who."

In the description of results provided below, we group judges' guesses in the following way: all guesses with confidence levels of 3 and 4 were scored as either right guesses or wrong guesses. All guesses with confidence level 1 and 2, along with refusals to guess, were counted as uncertain.[4]

Whenever judges changed their level of confidence. They were prompted to explain why. The entire session was recorded.

Results

There were two "phases" to the experiment. In "Phase 1" we conducted a total of twenty-four runs, roughly split among the four possible conditions. Both the color blindness series and the perfect pitch series treated in isolation support the hypothesis: in each case there were more correct guesses in the identify condition than in the chance condition. Since the numbers involved are small, however, a level of statistical significance is reached ($p = 0.05$ in Fisher's exact test) only when we combine all 24 runs

---

4. The results would not differ much if we used all guesses in the right-wrong analysis and ignored confidence levels.

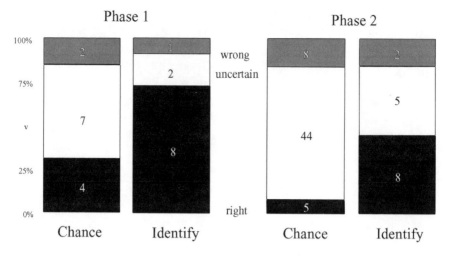

Figure 6. Results of proof of concept experiments

into just two groups—chance and identify.[5] Here we find that in the chance condition judges made 4 correct (high confidence) guesses out of 13 while in the identify condition they made 8 correct guesses out of 11.

We then carried out "Phase 2" of the experiment. This no longer involved interaction between judges and participants in real time. Instead, using ordinary email, we sent the transcripts of the recorded discourses to new judges possessing the target expertise to make new post hoc judgments. We may describe this in the following way: in Phase 1 of the experiment interrogator and judge were combined into one role; in Phase 2 the judge was not the interrogator. In Phase 2 the same dialogues could be sent to many judges and each judge could look at many dialogues. Again, the results of both modalities separately supported the hypothesis. In this phase the combined results were that 5 out of 57 (high confidence) guesses were correct in the chance condition, while the corresponding figure in the identify condition was 8 out of 15.[6] The results are represented graphically in figure 6, which shows the proportions

5. To use the Fisher test we have to construct a four-cell table in which wrong guesses and uncertain guesses are treated as one category to be compared with right guesses.

6. It is much harder to run the identify condition because two members of the esoteric group are needed—hence there are fewer identify condition runs.

of wrong guess (top band), uncertain guesses (middle band), and right guesses (lower band), for the combined modalities in Phases 1 and 2; the absolute numbers are also indicated.

When, as in Phase 1, the two modalities of Phase 2 were combined, the difference between the conditions was statistically significant (p = 0.028 in Fisher's exact test).[7] When the results of Phase 1 and Phase 2 were combined, the likelihood of the result being due to chance turned out to be p = 0.000 (Fisher's exact test).

We seem, then, to have shown that interactional expertise exists: the color-blind can pass as color perceivers relatively easily because they have been immersed all their lives in the language of color. In contrast, the pitch-blind cannot pass as pitch perceivers because they have not been so immersed.

Phase 3: Judges Have to Know What They Are Talking About

There remains a possible alternative interpretation of these results: that the difference lies in the judges not the participants. All judges possess the target expertise, and this means that the different conditions utilized different judges. It could be that being color-blind or having perfect pitch encourages one to think about the issues more deeply when compared with those who have "normal" abilities (i.e., they develop enhanced reflective ability). It could be, then, that identify-condition judges would be better discriminators in the chance condition too. That is to say, the judges' expertise might lie not so much in their knowledge of the domain but in their reflective ability. If it is reflective ability rather than domain knowledge that makes the difference, judges from a minority group would be better judges whomever they were judging.

To test this we had dialogues generated in the chance condition examined by the four most successful identify-condition judges, generating eighteen new guesses. In the new condition these four judges' discriminatory ability turned out to be no better than chance. This strongly suggests that the differences discovered in the main part of the experiment reported here were not primarily due to the differential reflective skills of the judges but to their understanding of the domain they were judging. In sum, judges who were good at judging "knew what they were talking about."

---

7. The results for each separate modality were also statistically significant in Phase 2.

Conclusions on Color Blindness and Perfect Pitch

One striking feature of the experiments was judges' lack of success in making correct discriminations in the chance condition. In Phase 2 the randomness of the outcome is especially striking, as can be seen in table 4, which gives the complete set of results for the chance conditions with the associated confidence levels ("0" indicates "cannot guess").

In the table a minus sign signifies a wrong guess, a plus sign signifies a correct guess, while "0" signifies inability to guess at all—the numbers indicate confidence levels. The rows in the table indicate the guesses made by different judges in respect of the same dialogue, while the columns indicate the performance of a single judge in respect of different dialogues. Simply by looking along the rows and down the columns one can see how random were the guesses in the chance condition. This shows that the color-blind are very good at passing as color-perceivers in the imitation game, while those with perfect pitch are very good at passing as those without perfect pitch. It does suggest, then, that a lifetime's immersion in the discourse of a group with a certain contributory expertise enables a person without the contributory expertise to acquire the corresponding interactional expertise, at least as tested by the imitation game, to a very high level—the strong interactional hypothesis is demonstrated. The contrast with the results of the identify condition reinforces the point and shows that without such experience it is hard to acquire interactional expertise in the absence of contributory expertise.

Table 4: Phase 2 chance condition guesses

Judges		1	2	3	4	5
	1	−1	+1	+2	−2	+1
	2	−2	0	−3	−2	−2
Color	3		+1	−2	−2	+2
Blindness	4	+2		+2	0	+2
Dialogues	5	0	+1			+2
	6	−2	0	+2	0	
	7	+2	−1	+1	+2	−3
	8	−2	+2		−3	+2

Judges		6	7	8	9	10	11
	1		-3	0	0	0	−3
Perfect	2	−2	+3		0	0	−3
Pitch	3	+1	+4			0	−4
Dialogues	4	−1	+3	+1			+4
	5	+3	−3	+1	−2	0	

Judges' Reasoning

The experimenters felt reassured that judges were making such identifications as they did make for "the right kind of reasons," but, although we asked for commentaries, there are few that are revealing. This is because judges tended to make their guesses on the basis of general impressions and feelings, often being the product of a build-up of small cues over a series of questions. As explained, judges were asked to give their confidence level on a four-level scale. The following five examples represent interesting interchanges in cases where a judge made a correct high-confidence-level guess. The first three extracts are from Phase 1 of the experiment, where the experimenters made notes of the judges' remarks; the second two extracts are from Phase 2, with judges writing their own comments. In each case the reader may consider whether they would have made the correct identification and how this relates to their own perceptual abilities (some grammatical and spelling corrections have been made for ease of reading):

1. A JUDGE WITH PERFECT PITCH
GUESSES CORRECTLY AT CONFIDENCE LEVEL 4

QUESTION 3: How do you pick up a single voice in a crowded room?

PARTICIPANT A: By hearing the individual pitch of the given person's voice, although it does depend on the volume of individual voices as well.

PARTICIPANT B: By trying to concentrate on that voice.

QUESTION 4: Would you rather use a score to arrange a piece of music or dictate by ear?

PARTICIPANT A: Dictate by ear.

PARTICIPANT B: Use a score, probably.

Here the judge thought that only a person with perfect pitch would use pitch to identify voices and that question 4 definitely indicated the person with perfect pitch. One can see that participants without perfect pitch would not know that these things could be accomplished by a person with perfect pitch and might very easily think that replying positively to the questions would be to fall into a trap set by the judge.

2. A JUDGE WITH PERFECT PITCH
GUESSES CORRECTLY AT CONFIDENCE LEVEL 3

QUESTION 3: Do you find perfect pitch a useful skill? Are there times that you wish that you didn't have it? Can you give me an everyday example of a pitch that is always the same?

PARTICIPANT A: Perfect pitch is a useful skill, though there are times when it is annoying. I hear the pitches in people's voices often so I can imitate quite well. An everyday example of pitch would be my computer—I know every "note" it makes as it boots up.

PARTICIPANT B: I have to say that it has become increasingly less useful as I have diversified the "styles" of music that I get involved with. It was useful when singing more traditional pieces, however I have been more interested in less formulaic styles in the last few years. I'm not sure that I understand what you mean by "everyday example." A pitch as a constant is a pitch at a constant.

Here the judge said that hearing pitches in people's voices was a valuable clue, especially since this was sometimes found to be annoying, whereas B didn't really seem to understand the question.

3. A COLOR-BLIND JUDGE GUESSES CORRECTLY, HAVING REACHED CONFIDENCE LEVEL 3 AT THIS POINT IN THE QUESTIONING

QUESTION 3: When shopping for clothes, how do you decide whether a tie will go with a particular shirt or jacket?

PARTICIPANT A: They have complimentary patterns.

PARTICIPANT B: I've never bought a tie in my life.

QUESTION 4: If not a tie how do you decide whether a shirt goes with a pair of trousers?

PARTICIPANT A: I follow the choices made by the shops' dummies or displays.

PARTICIPANT B: Just by intuition. If I'm having trouble I might ask my girlfriend. Otherwise, there are particular colors I like, such as red, which obviously goes with black.

QUESTION 5: What is the most irritating thing about being color-blind?

PARTICIPANT A: Having to do experiments! Not being able to follow conversations or other people's conversations when they discuss or mention particular colors.

PARTICIPANT B: When I was in school I drew a green squirrel—that was quite embarrassing! Generally it's not too bothersome.

Here the judge found A's answers to do with clothing implausible. He found the embarrassment caused by drawing something in the wrong color evocative of his own experience.

4. A COLOR-BLIND JUDGE
GUESSES CORRECTLY AT CONFIDENCE LEVEL 3

QUESTION 2: What colors do you have particular difficulty distinguishing?

PARTICIPANT A: Primary colors—reds, greens, yellows.

PARTICIPANT B: Greens and browns mostly, although there are others I have difficulty with.

JUDGE'S COMMENT: Participant A claims to have trouble distinguishing "primary" colors, whereas in my experience, it's the shades of color that present me with the most trouble; also why red, green, and yellow?!

In this comment we see that the judge is able to make a correct guess because his experience of what it is to be color-blind is not something shared by the pretender who does not know that it is shades that give most trouble. This contrasts with the facility with which color-blind persons were able to talk about the color perceivers' experiences.

5. A COLOR-BLIND JUDGE
GUESSES CORRECTLY AT CONFIDENCE LEVEL 4

QUESTION 2: What do you think are the main problems faced by color-blind people?

PARTICIPANT A: Functioning on an everyday level can be difficult—trying to get simple things done, e.g., identifying colored papers in a meeting, but also enjoyment of things can be affected—I might not be able to enjoy things like films as much as I can't see the same range of colors as other people, picking clothes in shops, that kind of thing.

PARTICIPANT B: Distinguishing between shades of color—this could make life difficult/interesting in a range of contexts—color coordination with regard to clothes, decorating etc. Obviously if color perception is extremely limited, this may impact on some kinds of occupational/leisure settings.

JUDGE'S COMMENT: . . . B's answer to the main problems one faces is almost exactly what I would say. . . . The real clincher for me may strike a color normal person as odd: the statement by A that "I might not be able to enjoy things like films." This seems a very strange idea, as I have never seen colors normally, so can't see how being color-blind would affect my enjoyment of them; although perhaps this reflects each person's personal outlook.

Again, the limited experience of the color-blind person's world by the color perceiver—e.g., what it means for the appreciation of films—means they are unable to reproduce the discourse of the color-blind.

Interactional Expertise and Science

Having "proved the concept" of interactional expertise and its relationship to immersion in a linguistic community, we tried the same procedure on one of the authors of this book who has spent years trying to acquire the interactional expertise of gravitational wave physicists.

This experiment is represented by the right hand circle in figure 5 above. The larger circle represents the majority of us—the "gravitational wave physics–blind" (gwB). The dark circle within it represents the small group of expert gravitational wave physics who can be said to be "gravitational wave physics–perceiving" (gwP). The very small white circle represents Collins's long expedition into the linguistic community of the gravitational wave physics perceivers. Collins has spent considerable time trying to acquire the interactional expertise pertaining to the field of gravitational wave detection.[8] Here we make no attempt to generate enough results to provide statistical significance, but once more we describe the judges' reasoning. This, then, is primarily a qualitative study, although it has quantitative aspects.

These tests were carried out in a much more simplified form than the proof of concept experiments described above. Participants were sent sets of seven questions by email about gravitational wave physics. In the first place the questions were composed by gravitational wave physicists who understood the purpose of the test. Respondents, one of whom was Collins, were asked to answer the questions from their existing stock of knowledge and understanding without referring to sources. The answers provided by Collins and by gravitational wave physicists were then presented side-by-side to other gravitational wave physicists acting as judges. Judges were sent a questionnaire to frame their judgments which also inquired about their academic specialism and experience. This was a Phase 2 type experiment with the generation of the questions being treated as a separate matter from the judging of the answers. Judges were asked to guess who was who without consulting any reference sources, and to provide the level of confidence in their guess on a four-point scale like that used in the proof of concept experiments. They were also asked to explain how they made their choices, if possible on a question-by-question basis. Table 5 shows examples of the questions and answers.

8. Collins has been conducting sociological studies on the area since 1975 with particularly intense exposure between 1995 and 2005.

Table 5: Some gravitational wave questions and answers

The Imitation Game For Gravitational Waves

Q2) Is a spherical resonant mass detector equally sensitive to radiation from all over the sky?

A2) Yes, unlike cylindrical bar detectors which are most sensitive to gravitational radiation coming from a direction perpendicular to the long axis.

B2) Yes, it is.

Q3) State if after a burst of gravitational waves pass by, a bar antenna continues to ring and mirrors of an interferometer continue to oscillate from their mean positions? (Only motion in the relevant frequency range is important.)

A3) Bars will continue to ring, but the mirrors in the interferometer will not continue to oscillate.

B3) Bars continue to ring; the separation of interferometer mirrors, however, follows the pattern of the wave in real time.

Q5) A theorist tells you that she has come up with a theory in which a circular ring of particles are displaced by gravitational waves so that the circular shape remains the same but the size oscillates about a mean size. Would it be possible to measure this effect using a laser interferometer?

A5) Yes, but you should analyze the sum of the strains in the two arms, rather than the difference. In fact, you don't even need two arms of an interferometer to detect gravitational waves, provided you can measure the round-trip light travel time along a single arm accurately enough to detect small changes in its length.

B5) It depends on the direction of the source. There will be no detectable signal if the source lies anywhere on the plane which passes through the center station and bisects the angle of the two arms. Otherwise there will be a signal, maximized when the source lies along one or other of the two arms.

Q6) Imagine that the end mirrors of an interferometer are equally but oppositely (electrically) charged. Could the result of a radio-wave incident on the interferometer be the same as that of a gravitational wave?

A6) In principle you could detect the passage of an electromagnetic wave, but the effect is different than for a gravitational wave. The reason is that unlike electromagnetic waves, gravitational waves produce quadrupolar deformations. A typical electromagnetic one wave would change the distance in only arm while a typical gravitational wave would change the distances (in opposite ways) in both, so the differential signal for the electromagnetic wave would be half that for a gravitational wave.

B6) Since gravitational waves change the shape of space-time and radio waves do not, the effect on an interferometer of radio waves can only be to mimic the effects of a gravitational wave, not reproduce them. An electromagnetic wave could, however, produce noise which could be mistaken for a gravitational wave under the circumstances described.

The main result is that judges who compared answers provided by Collins and answers provided by gw physicists were unable to identify the participants. If we count high confidence guesses (levels 3 and 4) as right or wrong and low confidence guesses (levels 1 and 2) as indicating uncertainty, we find that out of nine judges, two chose Collins as the gw physicist and seven were unsure.[9] This outcome was only possible because Collins did not make any technical mistakes in his answers. Collins, then, had demonstrated his interactional expertise according to the standards of the test. (Readers might like to make their own guess about who was who before reading on.)

Reasons that judges gave for their choice can be divided up into two classes. First, there are reasons based on the technical content of the answers. Some of Collins's answers were markedly different in technical content from the answers of the gw physicists. Second, there were reasons based on the style of the answers. Where technical content differed, it was thought that Collins's answers showed more evidence of being thought through in practical terms whereas the gw physicist's answers tended to be more theoretical or bore the marks of being drawn from a textbook. This was likely to be so because in one or two cases, such as question 5 in table 5, Collins had no choice but to think the answer through since he had not encountered anything like this question previously. Therefore, the "standard answer" did not come to mind (though he could readily understand it once he had seen it). Collins's answer, on the other hand, was correct so long as the question was taken to apply to the current and immediately foreseeable generations of detectors; the measurements needed to satisfy the more "theoretical" answer provided by the gw physicist are currently impossible to make. As one of the judges put it: "[X] has the much better answer. [Y] isn't entirely wrong, but his suggestion is impractical." Crucially, in such a technically demanding (and nerve-wracking) test, Collins's answers passed technical muster, and, given this, they had an air of authenticity in respect of contemporary experimental practice in gravitational wave physics. Here are examples of judges' reasons for preferring Collins's answers:

I find that I lean to [W]. But [Z] is pretty darn good. I'd be entirely unsurprised if you told me this was a control run and that you'd used responses from two experts.

9. Neglecting confidence levels and reporting on all the results, seven judges chose Collins as the gw physicist, one chose correctly, and one could not decide.

Set [P] looked more like they had been answered by looking up a book. Set [Q] looked as if they came rapidly out of the mind.

Some of the judges, and all of the judges in respect of some of the answers, did not feel that technical content allowed them to make a judgment. In these cases they fell back on style and, in the main, Collins's style was preferred because his answers were shorter and thus bore the hallmarks of someone who was answering impatiently—this suggested a scientist to other scientists.[10]

As a "control," we tried using persons who were not gravitational wave physicists as judges, comprising one scientist in another field and seven haphazardly chosen academics from the social sciences and philosophy. Again, counting guesses as right or wrong only if they were associated with the top two confidence levels, two of these nonspecialist judges correctly identified the gw scientist, one chose Collins, and five could not decide. If we ignore confidence levels, however, these judges' guesses were more successful, five being right, two wrong, and one unable to guess. Examination of the reasoning shows, unsurprisingly, that it was entirely based on style and in this case the judges tended to think that the more technical "text-booky" answers given by the scientist were the more authentic. This is perhaps best illustrated by the reasons offered by the two judges who made the incorrect identifications, preferring Collins:

j11: I have no idea about the detail of any sets of answers, not knowing this field. I thought [Collins] was more persuasive as he/she seemed not to feel the need to elaborate on answers quite so much or set them in some wider didactic context. As such, [Collins] did not strike me as someone trying to persuade anyone else of their own credentials, presumably because they are not in question.

j12: My guess was based on accumulating evidence from the series of questions, rather than any particular one. It seemed to me that the responses [of the scientist] were going out of their way to appear knowledgeable and "scientific/specialist." I suspect that the specialists actually talk to one another in more natural terms [as in Collins's answers], being able to assume shared background knowledge. I'm also aware, though, of how I'm interpreting responses to individual questions to fit in with my overall decision. As a possible get-out, of course individuals vary in manner—and

10. Collins is participant "B."

a very senior scientist might give different kinds of answers to a junior one than to another senior colleague.

Taken together with the tendency in the color blindness and perfect pitch experiments for judges sometimes to take long answers as indicating lying and sometimes short answers as indicating lying, we seem here to be exploring the literally understood "ethnomethodology" of decision-making in the imitation game.

To explore "the terrain" of the experiments further, we ran the experiment again in other configurations. Evans has a mixture of beer-mat and public understanding knowledge of gravitational wave science from reading Collins's papers, from talking to Collins over the years, and from helping to organize the first runs of these experiments. Evans, however, could not succeed in Collins's role in these tests. He could not pass as a gravitational wave physicist, being unable to use his knowledge in the creative way required. What he did know about gravitational wave physics is better seen as "information" rather than interactional expertise.[11]

We also asked physicists from other specialisms to try to pass themselves off as gravitational wave physicists, but they were easily detected because they made technical mistakes or showed glaring gaps in their technical knowledge. Collins also tried taking the role of both interrogator and judge and found he could use both questions he generated himself, and those set by other gravitational wave scientists, to distinguish gw physicists from other physical scientists who were not gw specialists and from Evans; he could do this without fail at confidence level 4 based on the technical mistakes or lacunae evident in the nonspecialists' answers.[12] The fact that other physical scientists in cognate fields (astrophysicists, astronomers, and relativists) could not pass the imitation game test is particularly interesting if one does accept that it is better if experts "know what they are talking about." The result should make us

11. Evans's knowledge of gw physics fits into one or other of the leftmost categories of expertise in the third row of the Periodic Table of Expertises.

12. Numbers were small for these last tests-only two runs in each configuration—but the technical errors and gaps were so obvious that it was pointless to go on.

Academics attending the 2005 meeting of the Society for Social Studies of Science (many of whom have science backgrounds) were also asked to try to judge between answers given by gravitational wave specialists and nonspecialist scientists. Of the 20 who replied, 8 guessed right (high confidence) and 2 wrong with 10 uncertain. (Unweighted results were 12 right, 7 wrong, and one could not guess). Again, the reasoning was based on the style of answers not the content.

still more concerned that in public life "scientists" are often asked to fill the role of generalized expert in matters remote from their specialism.

What we hope to have indicated is that the acquisition of interactional expertise can be detected and disentangled from less involving forms of knowledge with a simple experiment. Here again we have gone a little way toward showing that interactional expertise acquired deliberately in this way is much nearer to the richness of the interactional expertise of the color-blind than to even primary source knowledge. We have come a little closer to proving the principle of the strong interactional hypothesis.

The Role of Mathematics

Those composing the questions for the gravitational wave imitation games were asked to avoid mathematics. Were calculation or algebraic manipulation needed to answer the questions, Collins would have had no chance of passing the test. The banning of mathematical questions is less damaging for the implications of the test than might be thought because mathematics, as will be argued at greater length elsewhere, is not used much in ordinary discourse among scientists. Mathematics is integral to the physical sciences but only in the way that experiment is integral. Experiments are not being done when physicists sit down to talk about physics or make judgments about who to give grants to, and neither is mathematics. It is not the case that physicists talking physics over lunch talk mathematics, and it has to be case when, say, grant allocation committees meet because they are composed of scientists who do not know the field they are judging to the level of its algebraic or calculational niceties. For example, to know whether to fund a new gravitational wave detector, one might like to know how sensitive it will be in various directions: will it look at only a narrow solid angle of the sky or will it see waves from all around it? To work out the exact pattern of its sensitivity is a matter of mathematical analysis and calculation such as could only be done by someone with contributory expertise, but to see and understand a diagram representing the results of those calculations, and to understand their significance for the astronomical value of the instrument under review, does not require that the calculation be repeated, only that it is known to have been broadly agreed to be correct. In the light of this it is less surprising to find that some physicists do not use mathematics at all in their work. The appropriate model is a division of labor, where some physicists do the high level mathematical labor on behalf of their colleagues. Thus, to have interactional expertise in a science like physics does not require that mathematics be part of the talk.[13]

Overall Conclusion

The color blindness and perfect pitch experiments reveal the effect of linguistic socialization in extreme or "ideal" cases. Insofar as they can be taken to show that judges were making correct discriminations on the basis of tacit understandings rather than propositional knowledge, the experiments help to establish the idea of interactional expertise. They support the claim made above, that linguistic socialization has a larger role as compared to embodiment than would naturally follow from philosophical treatments such as that of Dreyfus. *Inter alia*, these experiments indirectly support the methodological claims of the assiduous participant observer. To establish the philosophical point still better, however, we would need to separate the different kinds of clues used by the judge in the imitation game. Some kinds of clues are of a "general knowledge" flavor. For example, any color perceiver would now know, having read this chapter, that one way to pretend to be color-blind is to make up a story about drawing some familiar object in the wrong color in one's early school years. That is the kind of propositional knowledge that could easily be encoded into a machine such as Lenat's CYC, Searle's Chinese Room, or any of the other imagined scenarios that depend on exhaustive knowledge of potential conversational turns in any one language. It is a kind of knowledge that belongs in the left-hand division of the specialist expertise row of the periodic table.

A more subtle procedure would concentrate on questions that explore the tacit components of socialization. One way of approaching this is to force the pretender to "repair" subtle mistakes in the judges' conversational turns. In respect of the Turing test proper, it has been suggested that the computer pretending to be a human might be given the task of editing passages of written text as in the "editing test" described in the coda to this chapter. Questions that demanded continuing repair of this kind could, perhaps, distinguish between the formal aspects of the knowledge of the participants and their more tacit linguistic abilities.

13. This argument is worked out at length with empirical confirmation in Collins 2008 (forthcoming). An anonymous reader of the draft manuscript asked a very interesting question to the effect "How do physicists setting the questions know what counts as a legitimate question in the test and what does not?" This would make an interesting topic for investigation in itself.

We hope that more imitation-game experiments with various protocols will be tried so that the test can be developed to the point where it can detect quite subtle differences in expertise. For example, perhaps the equivalent of the editing test could be used to tell the difference between those with beer-mat and primary source knowledge, on the one hand, and those with interactional expertise, on the other.

Coda: The Editing Test

(Reproduced, with permission of MIT Press, from Collins 1996b, 318–20.)

To make the idea of socialization more concrete I propose a . . . simplification of the "Turing test." Careful analysis of the test shows that the most difficult thing for a computer to do in such a test would be to make sense of badly typed or misspelled input. In other words, the really hard thing is subediting. Consider the following passage which is in need of subediting.[14]

MARY: The next thing I want you to do is spell a word that means a religious ceremony.

JOHN: You mean rite. Do you want me to spell it out loud?

MARY: No, I want you to write it.

JOHN: I'm tired. All you ever want me to do is write, write, write.

MARY: That's unfair, I just want you to write, write, write.

JOHN: OK, I'll write, write.

MARY: Write.

That we have the ability to correct passages of printed English of this kind is not a result of our fixed store of knowledge nor of a prolonged period of "training." It is not even that when we are being trained to do it we are corrected when we go wrong at each newly encountered example of such a problem; it cannot be this, because the new instances are equally new to any potential trainer and we merely beg the question of how the trainer knows what a reasonable answer would look like. Nevertheless, we usually come up with an acceptable version of even newly encountered problems of this type the first time we see them. We can do this in spite of the fact that what counts as an appropriate response—and there may be several possibilities—varies from place to place and time to time. This is

14. Thanks to Bart Simon for inventing the actual passage of text.

not surprising, as what counts as reasonable use of language changes as societies change. The ability to subedit reasonably successfully, then, is a matter not of learning a set of rules but of being a member of a society.

What I am proposing as a test of socialization is a comparison of the ability of fully socialized members of society, and those who might or might not be members, to subedit passages of this sort. To put this test into practice, the judge would have to be a full member of the society in question and would have to make judgments while being unaware of which passage had been edited by which entity. Like the original Turing test, the ability to separate members of a society from nonmembers using this test is a matter of probability, not certainty. (It is worth noting for the combinatorily inclined that a look-up table *exhaustively* listing all corrected passages of about the above length—300 characters—including those for which the most appropriate response would be "I can't correct that," would contain 10^{600} entries, compared to the, roughly, 10^{125} particles in the universe. The number of potentially correctible passages would be very much smaller of course but, I would guess, would still be beyond the bounds of brute strength methods.)

The editing test is much simpler to conduct than the Turing test and at the same time is a much better test for the abilities of computers in relationship to humans. Like the Turing test, it is a general test of the ability of one type of thing or person to imitate the actions of another type of thing or person.

CHAPTER FIVE

New Demarcation Criteria

In this chapter we return to the larger problem that we began with: Who should contribute to which aspects of technological debate in the public domain? At the start of the twenty-first century it is well established that the public should contribute to some aspects of these debates. The public have the political right to contribute, and without their contribution technological developments will be distrusted and perhaps resisted. This is what we called the "Problem of Legitimacy." Our complaint is that the social sciences of the last decades have concentrated too hard on the Problem of Legitimacy to the exclusion of other questions. As explained, our principal aim is to offer some way into what we call the "Problem of Extension." The Problem of Extension is concerned with how we *set boundaries* around the legitimate contribution of the general public to the technical part of technical debates.

We began our argument with the most commonsensical claim we could think of: only those who "know what they are talking about" should contribute to the technical part of technical debates. It is probably worth reiterating that this approach looks two ways. On the one hand, it is a conservative approach in that it restricts participation in the technical aspects of technical debates. On the other hand, it is a liberal approach in that it admits to the company of those who know what they are talking about many experience-based experts whose contribution would not have been countenanced in earlier times. It opens to the door to Wynne's sheep farmers and to Epstein's AIDS sufferers and, perhaps, to anyone with interactional expertise and, to a greater or lesser extent that is yet to be fully worked out, to various other groupings represented on the Periodic Table. Crucially, under this treatment, formal scientific training and accreditation are not the keys to the right to contribute—not even to the technical part of a technical debate. Under this treatment there is, then, a much narrower envelope of technical experts than under the folk wisdom view. In some respects there is even a narrower envelope than under the old authoritarian model of science, where any scientists was licensed to speak on any technical topic—we think only specialists of one sort or another should speak. On the other hand, there is a wider envelope of experts than under the old authoritarian model in that anyone with the right kind of experience, whether they have scientific training and accreditation or not, has a potential place inside it.

The bulk of the book has been to show how we might think about what it means to "know what you are talking about." Our efforts have concentrated on characterizing expertise and classifying it into its various types. We have concentrated heavily on interactional expertise because it is important and, so far as we know, it is a new idea that has not previously been examined. The philosophical discussion and the experiments on even interactional expertise are, however, merely a start. We and others are planning more experiments on the topic, and other scholars are critically examining the significance of the experiments and the idea of interactional expertise itself.[1] In other words, the discussion of interactional expertise found in this book is only the beginning of understanding of the concept. Furthermore, the analysis and experiments on interactional expertise reported here and continuing elsewhere constitute, at best, a "demonstrator project." Each category of expertise needs analyzing and researching in similar or greater depth.

Our approach to resolving the Problem of Extension, then, is to understand the nature and kinds of expertise. But were we to succeed completely in our aim, and our categories were widely accepted, and were a completely satisfactory way of using the categories to inform decisions to be found, the job would be far from complete. The reason is that something about the content of expertise as well as its forms has to be included in the exercise. This is because there are experts in every activity that takes place in social groups, from palm-reading to pop music. One may know *what* it is to know what you are talking about, and know *who* it is that knows what they are talking about, and know *how to balance* one way of knowing what you are talking about against another, but none of it helps if what is being

---

1. See www.cf.ac.uk/socsi/expertise for drafts and references to further work and see Collins 2008 (forthcoming).

talked about is the wrong thing. We have stated quite baldly that we think that knowing a relevant aspect of Western science should be a precondition for taking part in the technical aspect of a debate, but this begs the question of what is meant by "Western science." The previous chapters, and in particular the Periodic Table of Expertises, have explored what it means to "know" something. In this chapter we go where angels fear to tread and try to demarcate science and technology from other cultural endeavors. We make, in other words, another assault on the Problem of Demarcation.

Demarcating the Sciences from Nonsciences

The problem of demarcation is a philosophical standard. All existing attempts to demarcate science from nonscience seem to be logically flawed. For example, one would have liked to have said that science was marked out by the fact that its claims could be proved (or replicated), but the failure of logical positivism (and more recent work in the social studies of science) show that decisive proof is impossible (and that replication settles nothing where disputes run deep). Karl Popper suggested that the key should not be decisive proof but the possibility of decisive disproof: a hypothesis belonged to science only if the conditions under which it could be falsified could be set out. Lakatos's demonstration that it was as hard to prove something false as it was to prove something true uncovered the flaw. Importantly, despite failing to meet the strict standards of philosophy, these demarcation criteria still work well as argumentative tools; they encapsulate a lot of common sense about how a science should work.

It may be, however, that the philosophical standards are part of the problem and that this historical lack of success with demarcation criteria reflects a mistaken understanding of how social activities such as science should be understood. Wittgenstein showed that it was impossible to provide sharp defining criteria for the concept of a "game." Nevertheless, games can be recognized and things that are not games can also be identified, so why should the idea of science be any different? Like games, sciences can be recognized within the societies to which they belong, so "recognizing" cannot be the same as "defining an exhaustive set of rules for." This makes it possible to indicate what is meant by today's science, but only if the ambition of generating a timeless and universal set of criteria is set aside. In this chapter, then, we try to say something about the commonsense meaning of science. We try to develop a series of ever more narrowly specified demarcation criteria, always accepting that they will be flawed when compared to the standards of logic. We try to demarcate, first, science from art, second, science from politics, and third, science from pseudo-science. Only after this can we return to the role of expertise in the Problem of Extension.

Formative Intentions

To create new demarcation criteria, we are going to have to consider intention. The concept of intention has long been viewed with suspicion within philosophical circles because of the difficulty of knowing internal states. Since intentions are private, how can we identify them reliably?

The legal system demonstrates how hard it is to determine intentions. The courts utilize a complex adversarial system to try to tease out intentions and a fixed timetable to make sure that a potentially interminable argument is brought to a close, yet the courts still make mistakes. Fortunately, this is not our problem. Here we discuss only "formative intentions," which are quite different from intentions as traditionally discussed in philosophy and in courts of law.[2] Courts of law have to identify the internal state of some particular individual at some particular time—an "intention token." Formative intentions are the intentions that are available to actors within a form of life, and partly constitute that form of life, rather than being the intention of any particular individual at any particular time and place; they are "intention types."

Formative intentions are public because they are the property of the collective rather than the private property of the individual: they are available for inspection by anyone who shares the form of life which they help to constitute. To give an example, in the typically British form of life of the early part of the twenty-first century, it is possible to dance the Tango with a view to winning a Latin American dance medal but not to bring on rain. Readers of this book already know this; they do not need any special philosophical or legal apparatus to confirm it. This is not to say that there are no people living in Britain who might not be intending here and now to dance a Tango to bring on rain, only that their intentions are not constitutive of the British form of life and thus we do not need to worry about them; it is the form of life that interests us. We are concerned

2. Collins and Kusch 1998.

with the texture of social life, not biography or, to revert to the legal illustration, guilt. In our case we are concerned with the formative intentions that are part and parcel of the form of life of Western science. We can identify them because we already share that form of life. And when we rule things out of the science category we are going to be concerned with the formative intention types of the excluded groups, not the intention tokens of particular individuals.

Unfortunately, the first use of this approach is going to be weak and incomplete because many things that are not science share the intentional stance we describe. On the other hand, there are no things that are science that *do not* share this intentional stance. We begin, then, with a necessary but not sufficient defining criterion for science. As with other aspects of this project, the crucial thing is to make a start.

Science and Art

Consider the contrast between science and some of the avant-garde arts such as adventurous fine art, certain kinds of adventurous writing, adventurous music, and so forth.[3] The defining feature of these kinds of art is that, as far as most of their practitioners are concerned, they are intended, not to deliver information, but to engender often unanticipated reactions in the viewer. To give one example, the website of the Musée D'Art Moderne et Contemporain, in Geneva explains that "MAMCO aims to provoke people to reconsider their understanding of the notions of 'contemporary art' and 'museum.'"[4] Here the formative intentions are made explicit.

Let us call the enterprise signified by this kind of sentiment—the desire to provoke new and perhaps unique interpretations in the individual—"the provocative arts." Contrast this intentional stance with that of science. The intention behind writing a paper for a scientific journal has to be to explain the ideas or findings in the paper in the clearest possible way so as to afford only one interpretation to all readers—the universal interpretation. At least as a first approximation, should a reader take a different meaning from the paper than the author intended, then either the reader or the author has made a mistake.

Not that this is to say that the reader *can* take from the paper the exact

3. This argument was first presented by Collins at the 2000 Cardiff Millennial Quinquennial "Demarcation Socialised" conference.
4. http://karaart.com/swissart/museums/mamco/index.html.

meaning that the author intended—we know that even in the sciences, to a greater or lesser extent, "the author is dead." We know about "interpretative flexibility." We know, in other words, that the meaning of a paper once it passes from the hands of the author becomes, to some unavoidable extent, the prerogative of the reader. In science this is even institutionalized in the process of peer review and in what Merton called "organised scepticism." Nevertheless, irrespective of their ability to succeed in their aim, we can still talk about what the scientist-author and scientist-reader are *trying* to do. We can see that part of the very meaning of science includes the imperative that the scientist-writer should try to make the paper as clear as possible and as closed to alternative readings as possible and the reader of a scientific paper should be try to take from the paper exactly what the author thought he or she was putting into it even if they feel bound to criticize it after having understood it. One might say that though interpretative ambiguity is *intrinsic* to any sort of writing, in science it should never be *extrinsic*—it should never the goal of either author or reader. [5]

Turning back to provocative art, the *oughts* are different. If a "reader" takes away something radically different than what the author intended from an encounter with such a work, no-one has made a mistake. On the contrary, it may be that the greater the range of interpretations inspired in readers the better.

This difference has implications when it comes to the proper levels of expertise of those who *ought to judge* the works which come out of the two different forms of life. A critical comment on an early draft of a paper on expertise which was sent to us bears on the matter in an interesting way.[6] We were asked whether, in the case of a work of art such as Tracey Emin's unmade bed, notoriously displayed by the London's Tate Gallery in 1999, we would be happy to restrict judgments about its value and significance to a small group of people in the way that we thought appropriate in the esoteric sciences. What was interesting was the exact phrasing used in the rhetorical question:

> No-one without the training and exposure to appropriate gallery-going is . . . "competent" [to make a judgment]. So, can one derive the conclusion that only they *should* judge art?

---

5. The distinction between extrinsic and intrinsic is first introduced in Collins and Evans 2002.

6. Collins and Evans 2002. The next passage was also included in that paper. Steven Yearley kindly allowed us to identify him as the author of this comment.

Among other things, the commentator is drawing a contrast between an elite set of judges and the public at large and stressing the rights of the wider group to a legitimate and consequential opinion. The point is well taken because we all feel we have something to say on the artistic merit of "the bed." More significant, however, is that the elite group of judges was taken to be those with "training and exposure to appropriate gallery-going." This group of specialist observers, as opposed to artist-producers, was taken to be the legitimate location of elite expertise. So, the referee, in criticizing our undemocratic tendency to locate meaning in the sciences with the producers of knowledge, and in describing what he took to be the analogous tension in the arts, revealed something substantive about judgment. He showed that the "locus of legitimate interpretation" is different in the arts than the sciences. Put into our language, in the arts the competition between those who should count as suitable judges is between those with interactional expertise, on the one hand, and ordinary folk on the other, with the artists themselves—those with contributory expertise—"nowhere in sight." In science the competition is between those with contributory expertise and everyone else.

Art is made to be consumed, with provocative art being at the extreme end of the range, and that, perhaps, is why the locus of legitimate interpretation is where it is. If there is a judgment elite, it is made up of those with special viewing, or experiencing, expertise—expertise at consuming—not expertise at producing.

The folk-wisdom case—the case for the general public as the ultimate audience—is also much more easily made in the case of art than in the case of science. "I may not know much about art but I know what I like" is a less frivolous than "I may not know much about science but I know what I like." In the case of art we might be inclined to come down on the side of the skilled viewer as opposed to the public consumer, but at least the tension between lay and trained judges is easy to understand. Science, by its nature, is not directed at either kind of consumer but at the truth; this means that if we want to preserve it as we know it the audience should have less in the way of interpretative rights in respect to its meaning. Where the audience—as in the Mertonian norm of organized skepticism—does have rights, their first duty is to align their understanding with that of the producer, not to produce new meaning.[7]

---

7. It could be argued that at one time the public, or at least those who witnessed experiments, were more important to the process of science (see Shapin and Schaffer 1987).

The Chain of Meaning and the Locus of Legitimate Interpretation

Although we started with the provocative arts, we seem to have reached a conclusion that is also true, if in lesser measure, for a wider envelope of arts-type activities. If we think in terms of a "chain of meaning," with authors/artists putting meaning in at the left and readers/consumers taking it out on the right, we see that the legitimate locus of meaning, or locus of legitimate interpretation, is further away from the producers in the case of the arts than in the case of the sciences. How far this locus can move in the case of the arts is up for dispute. There are august institutions such as (in Britain) the Arts Council and less august institutions such as galleries and sponsors, there are the elite critics, and there is the general public. Then there are the postmodernist deconstructionists who believe the author loses all interpretative rights to the consumers/critics as soon as the work is "out of the door" (though whether these deconstructionist critics are thereby urging a democratic ideal or an elitist one—their own difficult to penetrate writings being the new elitist pathway into the arts—is unclear). But whichever way the argument goes, within the arts the locus is always somewhere to the right of the producers. The argument is represented in figure 7.

In figure 7 an "author" produces a work, which is interpreted by a member of the peer group, and/or a trained critic (or other institution), and/or the general public. The vertical arrow points to the widely accepted locus of legitimate interpretation (or what is widely accepted as a suitable intentional stance for the producer of the work), and can be positioned anywhere along the horizontal line. The argument is that where the arrow lies on the horizontal line depends on the kind of cultural activity in question. In the case of the hard sciences it is well to the left; in the case of provocative art, it is well to the right. In general, in the arts it is further to the right than in the sciences.[8]

Now let us try this idea out. It follows from the analysis that the role of journalists is likely to be different, and quite reasonably so, in the arts as compared to the sciences. In the sciences journalists rarely take it upon

8. There is one intriguing exception to the rule that begs for further analysis and research. This is the "transmuted expertise" of discrimination between experts which can occasionally be validly applied by the public (chapter 2). Although this does not feed back into scientific debate, it does appear to be a way of making sound technical judgments from the consumer end of the chain of meaning. It begs for a better understanding of the exact circumstances under which it could work reliably and invites a program of empirical research.

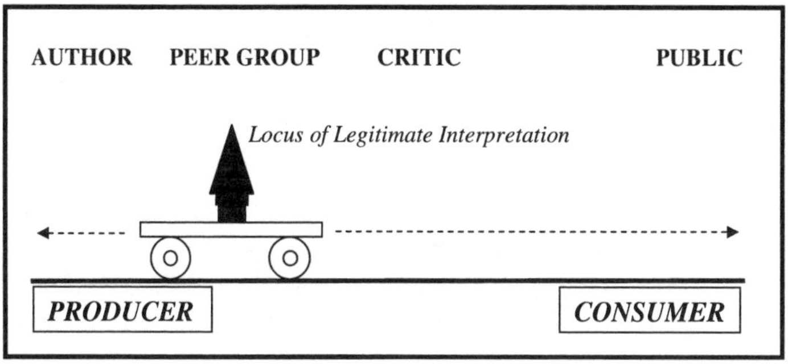

Figure 7. The chain of meaning

themselves to define the meaning of a work—usually they just sum-marize and describe.[9] In the case of the arts, however, journalists regu-larly try to define meaning. For example, many newspapers took their role in the case of "the bed" (and the notorious display, in 1972, of a simple arrangement of house bricks by Carl Andre) to be the champions of the general public's view that these were more like hoaxes than genu-ine works of art, and the artist was more of a Tony Hancock "rebel"-type than a Picasso-type, and we cannot but feel some sympathy for the claim. In the same way, moving a little to the left, it would not be unreason-able for a major newspaper's art or theater critic to claim to be making knowledge—legitimately defining what counts as a good or a bad play. In contrast, hardly ever would science journalists make the explicit claim that they were trying to make scientific knowledge when they wrote their stories.[10] The legitimate envelope of opinion-formers, and therefore the legitimate envelope of knowledge-makers, is far wider in the case of the arts than it is in the case of the sciences.[11]

9. For a view of the role of "science critic" which is rather different to the one taken here, see Ihde 1997.

10. With the exception of fringe sciences such as parapsychology where the normal boundaries are crossed (Collins and Pinch 1979). The normal boundaries were also crossed in the case of the UK reaction to MMR vaccinations, where journalists championed the view of the public at large on the basis of no scientific evidence. To put the argument another way, the distinction between constitutive and contingent forums discussed in Collins and Pinch 1979 does not apply to the arts because those in the "contingent" forum may legiti-mately *constitute* knowledge.

11. Literary criticism makes for an interesting case. It is all interactional expertise in respect of the literature itself yet it is also an expertise *sui generis*. For this reason literary criticism breaks the rule (discussed in chapter 1) that interactional expertise is parasitic—it can survive only so long as there is continued interaction with contributory experts. Here, interactional experts are the contributory experts in the expertise that they define with their

The idea of the chain of meaning also gives us a new term, "artism," which is the counterpart of "scientism."[12] Scientism (we are talking of scientism2), is the view that *every human activity* should conform to the science pattern of restricting legitimate interpretation to the author or those close the author at the left-hand end of the chain of meaning. Artism (artism2) is the view that every human activity should conform to the arts/humanities pattern where legitimate interpretation is always the prerogative of the consumer or someone close to the consumer at the right-hand end of the chain of meaning. Under scientism, the aspiration is that there should be no shifting ground anywhere and no attempt to create shifting ground. Under artism, the aspiration is that there should be no fixed points anywhere and no attempt to create fixed points—that is, there should be no knowledge claims which are not subject to legitimate reinterpretation by the consumer.

We can also use the analysis to shed some more light on hoaxes and fakes. An art hoax, or art forgery, or the like, should be easier to accomplish than a scientific hoax or forgery because its target can be a much less expert range of persons. It follows that such a thing is, and should be, much less of a scandal. Given that avant-garde art (as in the example of *The Rebel*, discussed in chapter 2) is presented to a wide public who are entitled to interpret it as they will, it is no surprise that such a thing might be pulled off. Instead of it being seen as a worrisome problem, it can be used as a resource for exploring the nature of art.[13]

Science, Technology, and the Locus of Interpretation

Having started with the provocative arts and moved inward to the arts as a whole, let us now move further left still and reenter the realm of the sciences and technologies. In "public use technologies," such as cars,

---

activities. This tendency seems to have been brought to a fine pitch in the postmodern style of literary criticism. Here, the content of the work of literature that gives rise to the critical discussion is almost irrelevant, and it would be unimaginable to bring in the artist or producer of work for an opinion, the critical dialogue taking place entirely between the critics.

On the same lines, a reader of an earlier draft raised the interesting question about whether there was an interactional expertise pertaining to language itself which would be different to contributory expertise in the language—speaking it. Our first inclination is to say that the two are identical, but this may be a point for further debate.

12. Presumably there are four kinds of artism just as there are four kinds of scientism as discussed in the introduction. There we said we hold with scientism4; we also hold with artism4.

13. Compare the case of the Sokal hoax and the Bogdanov debacle (chapter 2, note 16).

bicycles, computers, speech transcribers, and so forth, the public does have a legitimate right to interpret the meaning of the "works" produced by the "authors." Like works of art, public use technologies are made to be consumed, and thus the locus of meaning is further to the right than in the case of the esoteric sciences.[14]

This framework can also be used to understand some variations of practice even within the hardest of the sciences, such as physics. Here, however, the distinctions lie within a small range at the far left. The difference is between "evidential individualism" and "evidential collectivism."[15] It is between those who believe an entire scientific analysis should be completed by an individual, or within an individual laboratory, before it is released in the form of a publication and those who believe any data that is found should be reported early enough to allow other core experts to help in the determination of its meaning. This does not negate our initial claim about the intention of scientists, since it remains the duty of even evidential collectivists to be completely clear about what they are saying, and to be clear that the responsibility for making meaning remains within the core group of scientists and gets nowhere near the public. Following this line, we see that evidential collectivism can have real dangers when the rights to the interpretation of the data, having been exposed by publication, are taken to have moved beyond the core into the realms toward the right of the chain of meaning. This happens when scientists or others try to recruit the public in support of some theory or finding that is not yet accepted within the scientific community.[16]

Framing

The idea of the legitimate locus of interpretation is also a useful way of thinking about the notion of the "framing" of technological disputes. Consider the disposal of redundant oil rigs at sea, such as the Brent Spar platform in the North Sea. Analysts such as Wynne suggest that it is reasonable to frame the question in a way that has nothing to do with any particular oil rig and everything to do with society's willingness or otherwise to accept a certain *style* of disposal. Framed this way, the question is about the importance attached to preserving the environment in

14. See Bijker, Hughes, and Pinch 1987; Bijker 1995.
15. Collins 1998; 2004a, chap. 12.
16. This overextension of interpretative rights seems to be what happened in the case of the plane crash and the train crash discussed in Collins 1988 and in the case of the MMR vaccine.

general, and the decision is essentially about the kind of society we live in. Sinking the Brent Spar in the sea expresses a lack of respect for the environment—it is our respect for the environment that is the question rather than the actual pollution potential of the rig.

Under our analysis, Wynne and those like him are adopting a role in respect of technology like that of the art critic in respect of art. In art, the critic's role is to teach us how to interpret works of art—perhaps offering entirely different interpretations to those intended by the artist. The role is to create meaning from a position near to the consumer, perhaps representing the consumer. In the same way, the critics of technology shift the locus of meaning creation toward the consumer. For example, disposal of the Brent Spar platform is to be seen less as a problem to be solved at greater or lesser cost in terms of money and pollution—the way the disposers think about it—and more as something associated with lifestyle choice.

Seeing technology in this way may be legitimate in today's world, but considerations that belong at the left-hand end of the chain of meaning cannot be entirely ignored. Whether the Brent Spar would pollute the sea was germane to the debate about whether it should be disposed of at sea, even if it may not have been the decisive consideration. If it turned out that the Brent Spar would not pollute at all but rather would provide a safe haven for endangered species of fish, the same lifestyle choice might lead to a different conclusion. Whether it would pollute the sea or not, and whether it would provide a home for endangered fish or not, are questions with answers whose legitimate locus of interpretation is at or near the far left, not the right, of the chain of meaning.[17]

Here we are trying to provide a language and conceptual framework for thinking about who has legitimacy in such debates. If it is accepted, a consequence will be a slowing of the current slide to the right in the locus of legitimacy in the case of science and technology—toward artism2. It is no coincidence that the inspiration of some recent critiques of science is the European philosophy that underpins literary criticism; both movements are attempts to shift meaning (in terms of our diagram) to the right. Attempts to shift the locus to the right can be seen as another

---

17. This difference in emphasis maps onto the distinction between the technical and political phase of decision that was first introduced in Collins and Evans 2002. Although the word "phase" implies a temporal sequence, the usage owes more to the natural science meaning, where phase refers to the different states (solid, liquid, or gas) that a material might take depending on factors like temperature and pressure. In a similar way, the same decision might move between technical and political phases depending on the context.

move in the old war of the two cultures—they are attempts to treat scientific and technological knowledge as like knowledge belonging to the arts. With a shift to the right, critics, and the public in general, are given a role rather similar to that they would have in respect of the arts. But science and technology, if they are to retain their meaning as science and technology, must have a more authoritarian look about them because the locus of legitimate expertise will always be nearer to the point of origin of the knowledge. If, in our society, we want to retain the idea of Western science, we must want our scientists to want to be right—to be trying to stand on fixed ground—not just provocative and therefore happy with shifting ground.

Demarcating Science from Politics

Demarcating science from politics is easy once one accepts the idea of formative intentions. The problems of legitimacy and of extension arise because "the speed of politics is faster than the speed of science." Were there no distinction between science and politics, then they would run at the same speed because they would be co-extensive. Thus it must be possible to draw the line—the question is how.

We can start by drawing on the chain of meaning. One of the things we mean by "politics," when we contrast it with "science," is the mobilization of the interpretative power of those further to the right in the chain of meaning than would be normal in science as usually understood. In politics it is normal and appropriate for public opinion to be important in reaching conclusions; in science it is different, if not in practice at least in legitimate intention. For example, Shapin showed that, in nineteenth-century Edinburgh, scientists studying the brain were led to observe features that were homologous with their position in local Edinburgh politics and that the conclusions of the key scientists in the debate about brain structure were influenced by such local political considerations.[18] In our terminology, the influence of larger groups of townspeople effected the interpretations of the small groups peering at brain structures through microscopes. Other studies confirm that this was not a special case in that theory and observation alone must always show the influence of forces that are not normally thought of as strictly belonging to the activities of science. At the very least, the "small-p" politics of the scientific community is bound to enter into the formation of scientific consensus.

18. Shapin 1979.

This discovery has led some analysts to conclude that since politics cannot be removed from science, then science cannot be separated from politics. This is a mistake. The key, once more, is intentions. We do not say "Ah! Shapin's study has shown us the right way to do brain observations: get in as much local politics as possible." We do not say "the trouble with modern neuroscience is that local politicians have too little influence on the research findings." We do not conduct focus groups so as to sample public opinion more often and more accurately and thus help us better decide about the structure of the brain. We do not say and do such things because, although we know that all manner of "nonscientific" influences will effect the interpretation of what is seen through the microscope, we do not consider these to be "legitimate" influences. We admit that the locus of interpretation is further to the right than we once thought, but the locus of *legitimate* interpretation stays where it is. The intentional stance of science and scientists remains the same.

We can, then, demarcate science from politics, not by looking at the content of scientific knowledge but by looking at the contrasting formative intentions of scientists and politicians. To use the terms applied in the discussion of interpretative ambiguity, social studies of science may have shown that politics and other mundane influences are *intrinsic* to scientific knowledge, but, like interpretative ambiguity, they should never be *extrinsic*. Such influences must be resisted within any activity that we are to call a science. Those who engage in the social studies of science already know this, and that is why, while they proclaim on the one hand that science is invested with politics, they insist on the other hand that the testing of drugs is "unduly influenced" by the power of the drug companies, or that studies of the effects of smoking are "distorted" by powerful tobacco interests, or that genetics in the Soviet Union was "damaged," rather than "energized" by the political backing given to the ideas of Trofim Lysenko.

Demarcating Science from Pseudo-science

Demarcating science from pseudo-science is hard. At least some pseudo-scientists want to be clear and unambiguous in their work and to avoid the influence of mundane forces upon their findings; they wish to keep the locus of legitimate interpretation close to themselves. The extra criterion we have to introduce is the intention to make a body of work fit within the existing body of science. To make a start, we look at an extreme and somewhat colorful case.

Drug-induced State-specific Sciences

It has been argued that there are sciences that exist independently from the main body of science in perpetuity. An early proponent of this idea was the University of California at Davis psychologist Charles Tart. In 1972 Tart wrote a paper positing the existence of "state-specific sciences." Tart argued that there were sciences that were specific to altered states of consciousness such as were induced by drugs whose findings would have no bearing on the ordinary world of perception. These were the heady days of the counterculture, and the tenor of the times can be gauged from the fact that Tart's paper was published in the prestigious journal *Science*.

Tart's idea is germane for the following reason: if there were a science that was specific to the altered state brought about by, say, ingestion of LSD—let us call it LSDology—then it would follow that no-one who was not "on" LSD would understand its terms and categories. The idea of state-specific sciences suggests that a particular bodily state would be a necessary condition for the acquisition not only of contributory expertise but of interactional expertise too. If the availability of the concepts of a science (or any other activity) are tied to a specific bodily state, then even the interactional expertise pertaining to that activity cannot be acquired by those who are not in the state. Madeleine, fluent in respect of every other human activity according to Sacks, would remain dumb in respect of LSDology and, likewise, there could be no sociology of scientific knowledge of LSDology without LSD ingestion. A more complete and physical engagement with the form of life than talk alone provides— a physical engagement more appropriate to the acquisition of contributory expertise—would be needed. Sociologists should note, then, that the idea of a state-specific science puts sociology of scientific knowledge in jeopardy because once one allows that a level of physical immersion in a domain is an indispensable condition for the attainment of even inter- actional expertise, then practitioners can insist (as the "science warriors" used to) that if you cannot do the science you cannot talk about it.

State-specific sciences have no overlap, nor even potential overlap, with other sciences; the only people who will ever understand the state- specific sciences are those in the specific-state—otherwise they would not be state-specific sciences. They would just be a set of experiments or observations, carried out in specific conditions, the results of which could be communicated to the rest of the scientific community in the normal way. In the same way, state-specific sciences cannot draw on the findings of the main body of science, because these findings will not bear on the science of the special state. In a state-specific science there cannot be the trust for the findings of ordinary science that keeps most science going. Here we are introducing a new demarcation criterion that belongs within the formative intentions of science: "Except where specific new findings demand a break, the intentional stance of a science must be to maintain continuity as far as possible with the existing science."

Family Resemblance

We'll call this new rule "the family resemblance rule" after Wittgenstein's idea. As stated above, like other rules that try to express the normal way of going on in social life, it is vague: it very evidently does not contain the rules for its own application. We can try to find out what it means by exploring examples to which it can be applied, but first we will describe the idea of family resemblance.

Family resemblance implies that two distant members of a family, A and N, may have few characteristics in common but member A shares many characteristics with B who shares many characteristics with C, who shares many characteristics with D, and so forth, until N is reached. The idea can be represented diagrammatically as the set of overlapping ellipses depicted in figure 8. If the diagram represented the family of games, the very example that Wittgenstein used to develop the idea of family resemblance, the ellipse at the left-hand end might be soccer and that at the right-hand end might be dwile-flonking; if the diagram represented sciences, included among the ellipses would be, say, gravitational wave physics and ornithology.

The trouble with the idea of family resemblance treated as a logical rule is that certain family members may have nothing in common at all, so that having a family resemblance with science may rule out nothing.

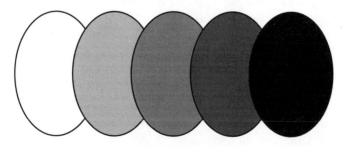

Figure 8. The idea of family resemblance

To make the idea work we have to invoke the idea of form of life to which it refers. Thus we know that when we say that both soccer and dwile-flonking are games, we know what we mean only because we share a form of life—the point being reinforced by the fact than many readers of this book, who do not share the appropriate form of life, will have to look up dwile-flonking on the Internet before they "get it."[19] It is clear that those who have to resort to the Internet are perfectly entitled to claim that dwile-flonking is not a game as far as they are concerned.

Intelligent Design

To use the idea of family resemblance, then, the family has to have some boundaries. This becomes clear if we examine the evidence given in the Dover School Board case in Harrisburg Pennsylvania in 2005. The case concerned the scientific status of the theory of intelligent design. The Dover School Board was prosecuted for allowing the teaching of intelligent design in science lessons whereas the complaining parents claimed it was a religious teaching. Giving evidence for the defense, a philosopher of science claimed that since Newton had been inspired by religion in his scientific work, and since it was unclear how science would unfold in the future, the religious motivation for the teaching of intelligent design could not disbar it from the category of science. In effect the philosopher claimed that religion was once central to activities that counted as belonging to the family of sciences and that one day it might do so again; religiously motivated scientific work could count, therefore, as having a family resemblance to science. A moment's thought will reveal that Newton was also an alchemist, and thus the same argument could be used for teaching alchemy in today's school science lessons. Furthermore, since

19. The meaning of game is brilliantly explored in the British Radio 4 ironic panel game "I'm Sorry I Haven't a Clue." The quintessential example within it is the game "Mornington Crescent." In Mornington Crescent each contestant in turn names a street or thoroughfare in London. It is clear from the long pauses for thought, the "oohs" and the "aahs," that an acceptable continuation that will cause difficulty for the other players is an accomplishment that can be managed with more or less wit and brilliance. After a series of such turns a player will suddenly announce "Mornington Crescent," to moans of disappointment, mild expletives, expressions of self-deprecation, etc., from the other contestants. It is clear that someone has made a mistake and the winner has cleverly exploited it. The viewers, and presumably the contestants, never have any idea of the rules. The joke is that they are playing the form of the game without the contents. Mornington Crescent is a "hollowed-out" game. Mornington Crescent, though it has the form of a game, is not a game but a comment on the meaning of game. Mornington Crescent, though it has the form of a game, is no more a game than Wittgenstein's discussion of games itself is a game.

we are in no position to know the future, the argument from what science might be like in the future could justify the teaching of anything that we have not so far thought of as science. Looking far into the past and far into the future, then, makes the notion of family resemblance unworkable. To use the family resemblance criterion we have to talk about science as we know it, not science as it once was or might be one day be.

It might be objected that family resemblance used this way is too conservative a criterion; it seems to say, "keep everything as it is." Once more we can refer to intentional stance. We know that science goes through stages of schism and revolution—what Kuhn called paradigm shifts. If one theory says that mass and energy are conserved while another says they are not, precipitating new ways of doing things in the laboratory and what amounts to a social revolution within science, does this not to create a rift in the family of existing sciences? The answer is "no" because the scientists pushing forward in the new direction have the intention to change as little as possible consistent with their new theories and findings. They do not want to overthrow the scientific method, nor the greater body of scientific findings, nor the major social institutions of science, nor the existing data of science. They do not want to become outsiders. Paradigm revolutionaries aim to persuade the same scientists to think and act in a new way within their existing institutions, preserving as much as possible of what already exists and their links with it. This contrasts with the proponents of ID, who want science to become much more compatible with something beyond science—the idea of a divine intelligence. Were they to succeed, the methods of science would change in that texts of obscure origin and the revealed certainties of faith would play a much larger role in the gathering of knowledge and the assessment of its value. The publicly accessible rules of science as we know it include careful observation at the expense of texts, especially those of obscure origin, and the elimination of personal bias such as comes with revelation. Furthermore, exploration of difficult topics, such as the development of complex organisms, would be discouraged, as a readymade explanation would be to hand, whereas current science demands that difficult problems should be seen as opportunities to deepen knowledge, the intention, even if it cannot be realized, being to understand and explain everything within science's purview by scientific means. All manner of additional rules could be added to this list, such as the centrality of experiment to the sciences. The crucial move we make here, in response to what has been discovered under "Wave Two of science studies," is to move away from taking these rules at their "epistemological face value." We locate their importance in the formative intentions that belong to the scientific form of life.[20]

Gravitational Waves, Parapsychology, and Astrology

Joseph Weber said he had discovered gravitational waves in the early 1970s, but his claims were rejected and he was treated as a maverick physicist for the next thirty years. His aim, however, was always to have his ideas accepted by the existing community of physicists; he published in physics journals, attended physics meetings, ran experiments readily recognizable as physics experiments, flawed or not, and continually tried to improve the sensitivity of his apparatus and his methods of analysis. He failed to convince anyone, but all these failures to convince were conducted within the recognizable form of life of physics. Weber's work is science under the family resemblance criterion; his findings may not have been accepted, but his intentional stance was to bring those findings into the ordinary work of contemporary physics.[21]

Now consider parapsychology: the study of telepathy, "mind over matter," "remote perception," and so forth. Parapsychology is a more difficult case because it claims to be able to show evidence for the existence of forces which most scientists believe do not exist. If these forces were real, their existence could cause an upheaval in the world of ordinary science since one implication might be that every "meter reading" would have to be safeguarded against the influence of the experimenter's and others' minds.[22] Nevertheless, if we concentrate on the small number of parapsychologists who work in university laboratories, we find that their aim is to demonstrate the existence of paranormal effects to the satisfaction of the contemporary scientific community, not to overturn science's methods or the larger part of its findings. The methods used by these parapsychologists are scientific to the point of tedium—very long controlled experiments supported by sophisticated statistical analysis. Parapsychologists occasionally publish in the peer reviewed journals of mainstream science, but they also maintain their own journals which operate demand-

20. For "Wave 2 of science studies," see Collins and Evans 2002. For the centrality of experiment, see Crease 2003.
21. Collins 2004a, esp. chap. 20.
22. A few scientists, however, do believe that quantum theory leaves a space for the effect of the observer's consciousness on the way events turn out at the macroscopic scale.

ing standards of technical review. Insofar as these parapsychologists want to overturn anything, they want to overturn the theoretical consensus about the elementary constituents of the universe—in this respect their project is like that of Einstein at the start of the twentieth century. Like Einstein, they hope to preserve as much as possible of the existing institutions and methods of science. Parapsychology is not, then, something that could *never* be a common or garden science. Parapsychology is, like cold fusion, a science that is not successful enough to have given rise to much overlap with current sciences but is still trying.[23]

If we now turn to newspaper astrology, we can see that almost none of the ideas, none of the institutions, and none of the personnel who endorse it have any overlap with orthodox science. What is more, the majority of astrologers do not want to have any such overlap—they do not want to become part of orthodox science. Astrology could go another way! If science ever were to accept the influence of the planets on human destiny and embrace the "astrologers," the astrologers would be very different from those of today and the influence of the planets would be mediated by forces within accepted science (or within the accepted science of the day). The forces, for example, might be the relationship between birth date and climate or even the self-expectations engendered by believers in astrology born under certain star signs (a kind of placebo effect such as is central to medical science). The aim of those endorsing such "astrologies" would be to *create* a family resemblance between the enterprise and science; they would want to be absorbed into mainstream science, perhaps changing the cognitive content of some sciences (probably psychology rather than astronomy) in the way Kuhn describes.[24]

The family resemblance criterion is a guideline. It says that the preservation of discontinuity between a new approach and the main body of science cannot be anyone's intention if the new approach is to be counted as science. There may be long-lasting conflicts, which turn out to be irresolvable differences of view, but this cannot be what scientists have in mind. In sum, under this view no-one can proudly proclaim that it is possible, within science, to believe both p and not-p. Studies of science

23. Once more, the social psychology of polarization forces us to add this footnote stating that we are not here supporting the claims of parapsychology.

24. Renee Gauquelin apparently discovered a correlation between star sign and athletic prowess. Gauquelin gets into science under our criteria even though he was rejected by the Committee for the Investigation of the Claims of the Paranormal. See the article by Gauquelin in *The Truth about Astrology*, trans. Sarah Matthews (Oxford: Blackwell, 1983) and John Anthony West, *The Case for Astrology* (New York: Viking Arkana, 1991).

over the last three decades have shown that it is in fact possible for science to hold both p and not-p—at least for periods of around half-a-century and longer—but this is not something to be valued.[25] To use our favored language, discontinuity is intrinsic to science but not *extrinsic*. Thus, alternative medical treatments which value the idea of treating the "whole person" *are not* science if the aim is to combat the existing body of medical science, or move medicine toward the realm of magic (even if this can work as a cure for the individual), but *are* science if the aim is one day to understand the whole mind and body as a complex system of causal chains.[26] Antivaccination campaigns *are not* science, although studies of the dangers and costs of vaccination campaigns compared to the dangers and costs of the diseases they are designed to combat *are* science. Even the claim that the complexity of certain organisms is too great to be explained by evolution *is* science if the object is to find better scientific explanations, but it *is not* science if the aim is to replace elements of science with elements of religion. In terms of the chain of meaning, the aim of scientists must be to return the locus of legitimate interpretation as far to the left as possible even if it sometimes takes excursions a little way to the right.

This chapter has been intended to bring us to the point where we could begin to use the Periodic Table of Expertises to help us make decisions about who counts as an expert and who does not in respect of the specifically technological aspects of technological disputes in the public domain. The Periodic Table in itself cannot demarcate sciences from the arts, from politics, and from pseudo-sciences, all of which have their own experts with their own expertises ranging from beer mat, to contributory, to meta-level. But if the new demarcation criteria work, or some other demarcation criteria can be made to work, they can be used to help identify scientific and technological experts as opposed to other kinds of experts. Once the domains of science and technology have been identified, an appropriate balance of experts and expertise from within the domains can then be made.

25. See, for example, Collins and Pinch 1993/1998.
26. Collins and Pinch 2005.

Science, the Citizen, and the Role of Social Science

To repeat a point we have made over and over, we are not claiming that there is nothing more to making technological decisions than sorting out the appropriate groups of experts. Technological decisions within the public domain have a political as well as a technical phase. There are three questions about the way the two phases relate to each other.

1. Given that there is an analytic distinction between science and politics, what is the appropriate ratio of science to nonscience in a decision?
2. Given that there is a distinction between science and pseudo-science, which are the sciences proper that should bear upon a decision?
3. How can the public recognize what they need to recognize to make appropriate decisions?

Both 1 and 2 are components in the problem of "framing" a technological debate. As mentioned earlier in chapter 5, one good justification for an element of nonscience is the opinion of the consumer in cases of public-use technologies such as cars and personal computers.[1] Another justification for a large ratio of nonscience or nontechnology in a decision, be it politics, consumer preference, or lifestyle choice, is weak science. If

---

1. An interesting borderline case is market research among "lead users" of new technologies. In effect, this group of users acquire contributory expertise in the use of the programs, and the companies consult them as experts. Interestingly, the groups of lead users who contribute in this way may be too narrow. It has been argued that truly novice users can be neglected in consequence of dependence on lead users. The result is that the design of, for example, IT interfaces turns out to be less than optimum for the majority of users. See, e.g., Von-Hippel 1988 for lead users and the arguments of Phil Agre at http://commons.somewhere.com/rre/2000/RRE.notes.and.recommenda19.html.

scientific consensus lags a long way behind the need for public decisions or the formation of public preferences, the science should not have a big input. Sciences are of different kinds, from the exact physics of planetary motion to the intractable problems of long-term weather forecasting or economic modeling. Macro-economic forecasters try to predict the inflation rate or the unemployment rate of a country a year or so in advance. Their forecasts are based on masses of historical data, a first-class understanding of economics, and complex computer models, but they can still get it wrong, sometimes very wrong, and they often disagree markedly.[2] Does this mean they and their enterprise should be abandoned? Is this a case where there is nothing to go on but politics? Our view is that even here there is a case for nurturing and listening to the experts, however impotent they are in terms of the outcome of their forecasts.

The first reason for this is that economic forecasters simply know more about how the economy works than anyone else—they know what they are talking about. A more subtle argument can be drawn from the partly analogous case of art forgery discussed by Nelson Goodman.[3] Goodman asks why it is important to maintain the distinction between genuine works of art and fakes when even the most distinguished critics cannot tell which is which (except by forensic means, such as tracing the history or analyzing the paint, canvas, etc.). Goodman argues that appreciating art is a developing accomplishment and that even if today's art critic cannot tell the difference between the genuine item and the fake, the skill of reading paintings might one day develop to the point where the difference will become clear to the trained and practiced viewer. This can be guaranteed never to happen unless the institutional difference between the fake and the real is preserved along with that of the idea of the art connoisseur. In the same way, we can argue for the preservation of the institution of expert economists even if current economic prediction is often far from the mark—one day it may be possible to make more accurate predictions. And even if this will never be the case for the economics, it may be the case for other related expertises; hence, do not give up expert institutions lightly even if they are currently of little use.

That said, experts should obviously have a relatively greater input where their results are more reliable. If economic prediction were a more certain science it would carry more weight, even though the decisions

2. Evans 1999.
3. Goodman 1969.

that invoke economics will always be political decisions. Where we move further from politics, the weightings will change again. The public panic in Britain over the safety of the combined mumps, measles, and rubella (MMR) vaccine, which surfaced in the early 2000s, seems to be a case where the weighting of expertise went wrong. With no real controversy within the scientific community, the value of consensual expert advice was entirely misjudged by journalists, the public, and, unfortunately, many social scientists.[4]

At the same time, it is rare to find a technological decision that calls for nothing more than the opinions of specialists. None of the argument in this book should be taken to mean that in many or most technological decisions certain groups of experts do not try to aggregate unjustified power to themselves. Cost-benefit analysts tend to work with too narrow a range of easily measured costs; risk analysts do the same, not accounting for such things as the political risks associated with safeguarding nuclear installations against terrorism. Neither group can account for the cost or risk of loss of freedom against the cost of a reduction of CO_2 emissions. Clearly, these are elements of the decision that fall squarely into the nonscientific domain.

More subtly, even where we confine the discussion to esoteric domains, what we have shown is that scientific and technological expertise can often be found outside the narrow confines of the accredited scientific and technological community. When he introduced the term "tacit knowledge," Polanyi was trying to establish the need for a "Republic of Science."[5] Scientists must rule themselves, he said, because, being the only possessors of the inexplicable tacit knowledge, they were the only ones who understood their capabilities and purposes. Implicit is the idea that only practitioners are fit to judge other practitioners. In our terms, this would amount to the claim that only those with contributory expertise can judge others with contributory expertise. The trouble with this view is that we now know that even within the scientific community acceptable judgments are often made by those with no more than interactional expertise (and sometimes referred expertise) in respect to what they are judging. If interactional expertise has such an important role within science, it opens the door to the possibility that nonscientists with interactional expertise in a science might also play a role in judging it. There are also unaccredited contributory experts, such as Wynne's sheep

4. Speers and Lewis 2003, 2004; Collins and Pinch 2005.
5. E.g., Polanyi 1958.

farmer's or Epstein's AIDS experts. Thus the walls of Polanyi's Republic are breached even as the special value of expertise is recognized.

On the other hand, social scientists can all too easily fall into the trap of thinking that those outside the Republic are making technical choices whenever they make a choice about a technical matter. If political choices and lifestyle choices were assimilated with technical choices, then the form of life we call science and technology would disappear. Let us return to the quotation supporting the folk wisdom view in the matter of *The Politics of GM Food*, first presented in the Introduction:

> ". . . many of the public, far from requiring a better understanding of science, are well informed about scientific advance and new technologies and highly sophisticated in their thinking on the issues. Many 'ordinary' people demonstrate a thorough grasp of issues such as uncertainty: if anything, the public are ahead of many scientists and policy advisors in their instinctive feeling for a need to act in a precautionary way."[6]

A closer examination shows that this claim was based on the outcome of a series of focus group meetings, the discussions being well documented in the report. The authors explain that "biotechnology was too scientific for people to comprehend easily and many wanted to know more about what was involved" (7). The focus group members felt lost in the technology: "It's all very technical to me" (40); most people were "confused about the process of cheese making and the role of the enzymes, calves and so on" (45); "You're not aware because you haven't been educated about biotechnology and the fact that it is actually happening. How many people who you ask in the street about 'do you know about biotechnology in food?'—they'd say 'no'"; "'Genetically modified'—Most people wouldn't know what that meant"; "I think the main worry is that we just don't know how these things have been modified or what they have done to make it replace that"; "What they've actually done to the tomato. Yeah what they have actually done? In simple terms for the average housewife shopper?" (52).

On the other hand, the focus group members' distrust of the institutions of science, technology, and government seemed more coherent, and given their experience of successive British governments' mishandling of these issues—which gave them something from which to generalize—this is understandable. What remains worrying is the way the authors of the

---

6. *The Politics of GM Food: Risk, Science and Public Trust* (Swindon: Economic and Social Research Council. ESRC Special Briefing No. 5, October 1999), quoted at page 4.

report unashamedly assimilate moral choice and choice of lifestyle with technical judgment. "In general, the anxieties were more pronounced the closer particular proposals came to challenging people's sense of an established moral order" (16–17). "There was a feeling that scientific and technical interventions moved food even further from its (desirable) natural state" (7). The distrust of GMOs was also based on an "instinctual" (the word is used approvingly in the report) distrust of the "unnatural." This is documented on page after page of quotations from focus group members in an annex to the report (37) as well as in the report's analytic main body. But trustworthiness of the natural is almost a paradigm of nontechnical reasoning. The Irish potato famine was "natural," the epidemics of plague and smallpox that killed our ancestors were natural, AIDS is natural. On the other hand electricity is unnatural, penicillin is unnatural, and, pertinently, miscegenation is seen by many as unnatural, although this attribution is now resisted in what many of us would call civilized societies. The argument from the natural is about as unsophisticated an argument as one can find, yet it is quoted in this report as supporting the view that the public exhibited a "reasonable" assessment of the evidence.

Among other things, it is this kind of confusion of the technical and nontechnical that we want to clarify; we want to separate the technical from the political phase of the debate. The report reveals that the public's technical knowledge about GM was poor—the very most they had was a vestigial "transmuted expertise" which arose from their reasonable distrust of official spokespersons in respect of the new technology. But in spite of this lack of technical knowledge, the public's right to demand that new GM technologies be tightly regulated remains unaffected. What we should be celebrating is this political right in a democratic society, not the spurious technical abilities of the public.

Even while exploring and defending the notion of specialist expertise, we have argued that the public at large has rights in the matter of technological decision-making. They have the right to choose their politics, their lifestyles, the risks they take, and even the extent to which they trust scientists and technologists. The social science of the last three decades has provided an intellectual grounding for hugely enhancing the rights of citizens in these respects. It has done it by "leveling the epistemological playing field." Science and technology no longer stand so far above the common terrain of knowledge that they seem to belong to some heavenly domain—a domain of revealed knowledge and priestly authority. Science and technology have become ever more familiar, their ways of assessing evidence bearing a disturbing resemblance to everyone else's ways of assessing evidence. This demystifying of science and technology has been entirely a good thing and has reduced the chance of the growth of a new kind of science fascism built on foundations of quasi-divine infallibility, which, in the 1950s, seemed a real possibility. Scientists and technologists, we have learned, cannot displace politics with expertise.

Nevertheless, to take it that the epistemological landscape is without a vertical dimension is to abandon responsibility for the world we live in. The new job of social scientists, having been so successful with the leveling, is to rebuild some structure—or, more properly, since it is obvious that there is lots of vertical structure—to understand what holds things up. What makes it that, though there are no tall mountains left, there is also not just liquid mud; what makes it that, at the very least, there are pretty substantial hills.[7] We are suggesting that the answer is expertise and experience and that the academic study of expertise and experience (SEE) is one possible new role for those who take epistemology seriously. The leveling down has resolved the Problem of Legitimacy; the study of the remaining hills may resolve the Problem of Extension.

Given SEE, how can the public, with their rights to make political choices intact, best bolster them with wise decisions with respect to the purely technical part of technical judgments? In the light of our analysis, the answer appears to be that, in the absence of suitable specialist experience, the citizen can make technical judgments only through the transmutation of expertise that starts with the social expertise of ubiquitous and local discrimination—a matter of choosing *who* to believe rather than *what* to believe. Surely, one of the tasks of the social sciences is to help the citizen make better discriminations of this kind by revealing more fully the social processes of science and by explaining the kinds of expertises that bear directly on matters of science and technology. If those quasi-technical judgments open to the technically inexperienced citizen are to be based on social judgments, then the better the understanding of the social processes of science, the better the judgments are likely to be. Transmuted knowledge does not make the citizen a scientific expert capable of contributing to the question of whether it is "p" or "not-p" that is true in any particular scientific debate, but it can help the

---

7. The levelling of mountains metaphor is taken from page 141 of Collins and Pinch's *The Golem* (1993/1998). The book has sometimes been hugely over-read as justifying a complete levelling, even though just a couple of pages after the metaphor it is said that scientists must be appreciated as the best available experts in the natural world.

citizen make a sensible decision about whether his or her political decision should be premised on p or not-p.

Like it or not, those who study knowledge are experts in the nature of knowledge. If we refuse to acknowledge any role other than criticism—if we are willing only to level down and never to build, explain, or evaluate the structure of the vertical dimension of epistemology—we are evading a responsibility that only we can fulfill. This is to bring an understanding of the social dimension of knowledge to bear upon the ways that knowledge is used. We must, then, be ready to explain science, and explain the nature of expertise, to as wide an audience as possible.

We must also be ready to explain expertise and assess expertise where more specialized audiences are concerned. Social scientists, philosophers, and other experts on expertise must be not just ready but anxious to offer advice on the nature of experts and expertise wherever expertise is used. And this means more than simply pointing to the contrast between scientific knowledge under the canonical model and scientific knowledge in its more social guise; it means more than pointing out that certain judgments about who the experts were can only be made with hindsight; it means more than pointing to the tension between the idea of expertise and the idea of democracy. It means working out some way of deciding how to use expertise even when we know it is much less sure than once we thought it was, even when we know it is too early to know who the experts really are, and even when we know that it seems undemocratic to select a group of experts, however wide, to whom we grant more authority than we grant to the ordinary citizen. We must be ready to alleviate the tension between democracy and expertise by helping with the design of citizens' juries and consensus conferences: helping not just by saying "let us bring in some citizens" but by stating what kinds of citizens with what backgrounds would be best and what kind and what length of exposure to what sort of technical material might turn them into better representatives of the rest.

Finally, if any of this has been convincing, we need more research on expertise. We need more research on interactional expertise to find out just how much practical experience can be excluded while still attaining interactional expertise good enough to make sound judgments. Could Madeleine (chapter 3) ever become an interactional expert in gravitational wave physics? One can start to answer the question by asking whether the congenitally blind are just as capable of passing as color perceivers as those who are merely color-blind. One might even ask (and we and others hope to do it), whether the same areas of the brain appear to be utilized by a congenitally blind person talking fluently about the color red and a normally sighted person talking fluently about the color red. This is not to prejudge what the answer to such an empirical question might mean, but how can one not want to know the answer if one is interested in the way knowledge is embedded in language and society?

We need research of similar depth into the other categories of expertise. In just what different kinds of ways would someone with primary source knowledge, or popular understanding of science, fail the imitation game test? How do trained and qualified but nonspecialist scientists fail in comparison with untrained and unqualified persons with specialist primary source knowledge. (We already know from chapter 4 that trained and qualified nonspecialist physicists perform less well in imitation game tests than a specialist with only interactional expertise.)

How do connoisseurs compare with contributory and interactional experts? What exactly is referred expertise and how does it work? How reliable is discrimination—the transmuted expertise—and how might we test it for reliability? How reliable is social understanding of science and its context as a source of transmuted knowledge and how might we test it, or even improve it? Could it even be the case that where expertise is especially poor at making predictions—let us say about next year's weather—we should replace it with a vote? We have argued that we should not but, no doubt, this argument needs more study and analysis. What is sure is that we cannot imagine any outcome of such an analysis that says that the vote should play a part in determining expert weather forecasters' predictions about next year's weather and their understanding of its causes, even if a vote should play a crucial role in determining policies that turn on predicted levels of sun and rain.[8]

We believe that it is the job of those sections of the social science and philosophy community who study knowledge to know what expertise is and what are its types and levels. The bulk of the book has been an attempt to begin to answer these questions. We have put together a candidate "Periodic Table of Expertises" based on the notion of socially located domains of tacit knowledge. The enterprise has been intended to establish that even in the face of the new Weltanschauung, with its distrust of science, expertise can be identified independent of its social attribution. We have argued and inscribed in the table that the location of the boundaries

8. Surowiecki (2004) argues that a democratic vote about the future rate of inflation is likely to be significantly more accurate than the predictions of economists. His argument is badly in need of further study, however, currently being suspiciously anecdotal.

of real expertise are not coextensive with the boundaries of accredited expertise; there are kinds of expertise that are not captured by traditional modes of accreditation. These boundaries include within them what have been misleadingly called "lay experts"—who should have been called "experience-based experts." They also include interactional experts—those who are expert in the language of a specialist domain if not in its practices. Since most decisions even within science are made through the medium of interactional expertise, it must be counted as a high level of expertise when it comes to decision-making. These high level expertises do not exhaust the domain of technical decision-making—there are lower level expertises that bear upon it too. How these should be balanced in any decision is a matter for further analysis and research. The crucial thing is that this further analysis and research treat expertise as real. Only this way can the social sciences and philosophy contribute something positive to the resolution of the dilemmas that face us here and now.

Waves of Science Studies

In "The Third Wave of Science Studies: Studies of Expertise and Experience," a much discussed, some would say "notorious," paper published in 2002, the authors used the device, "three waves of science studies," to capture the state of the discipline. Very roughly, the first wave was that in which the problem for disciplines that study science from the outside was to explain science's success and to work out how to maintain the conditions for its success. Under this wave it would have seemed odd to ask whether science was successful—following the contribution of science and technology to the Second World War, the success of science was a premise. Under Wave One, authority in matters of science and technology naturally flowed from the top down, with the scientists and technologists at the top.

The second wave began in the 1960s (though it had notable precursors).[1] With hindsight it can be seen as a reaction to the first wave. Scientific and technological knowledge were now more likely to be described as "social constructions," and the grounds for their epistemological priority were questioned and swept away. Relational theories of knowledge held sway. The authors of this book made significant contributions to this second wave.

Studies of Expertise and Experience, the main topic of this book, were described in the 2002 paper as a third wave of science studies. The second wave had proved a fertile ground for the promulgation of the folk wisdom view, and the authors felt it important to challenge it with a nonrelational analysis of expertise. The paper called for a realist treatment of expertise that would provide for a more systematic analysis of normative

---

1. Especially Fleck 1935/1979.

judgments about who had expertise and who had not. Recognizing the many contributions to science policy to which the second wave had given rise, the paper called for social scientists who studied science to start making their own judgments about expertise from upstream rather than merely studying others' attribution of expertise from downstream. It was recognized that all judgments were fallible, so this would mean exposure to new kinds of intellectual risk, but it was argued this should not to lead to paralysis. Fallibility does not lead to paralysis in the natural sciences, and it certainly does not lead to paralysis in politics, so it should not lead to risk avoidance in science studies. The claim was made that, although their judgments would be fallible, those who studied expertise "knew what they were talking about" in matters of expertise and had a major responsibility to use their expertise about expertise. We now summarize the differences between what we called Wave Three and the other two Waves of science studies.

Differences between Wave Three and Wave Two

1. *Upstream not downstream:* The aim of Wave Three is to change the world not just observe it.
2. *Insecurity:* As a consequence of being upstream, the claims made under Wave Three will be less secure than the claims made under Wave Two, just as science is less secure than skeptical philosophy.
3. *Categorization:* Both 1 and 2 imply that the analyst's reflex under Wave Three will be to construct categories rather than dissolve the boundaries between them.
4. *What is celebrated:* Under Wave Two what was celebrated and exploited was scientific uncertainty: scientific problems were turned into sociological resources. Under Wave Three scientific uncertainty is not a resource but gives rise to the question about how to act in the face of its inevitability.

Differences between Wave Three and Wave One

NB: Many of these are also differences between Wave Two and Wave One.
1. *The fifty year rule:* Scientific disputes take a long time to reach consensus, and thus there is not much scientific consensus about.
2. *The velocity rule:* Because of the fifty year rule, the speed of political decision-making is usually faster than the speed of scientific consensus formation, and thus science can play only a limited part in technological decision-making in the public domain.

3. *Uncertainty:* Even if the velocity rule does not hold and there is consensus among the scientific community, science is imprecise enough to make it likely that, under conditions of dispute, residual uncertainties will allow it to be "deconstructed" and disqualified from giving a firm guide for policy.

4. *Intrinsic not extrinsic politics:* All scientific decisions are intrinsically political, and that is another reason why they cannot form an unproblematic basis for political decision-making even when there is scientific consensus. Nevertheless, this does not mean that politics should be extrinsic to science.

5. *Punditry:* Scientists cannot speak with much authority at all outside their narrow field of specialization.

6. *Experience:* The major ground for judging expertise is experience, and it widens the base of expert decision-making beyond science and technology professionals.

7. *Fundamentalism:* Though scientific thinking is central to our form of life, it cannot form a basis for judging other kinds of thought such as religious, artistic, or romantic thought; scientists are to be treated not as authorities but as experts—plumbers not priests.

8. *Framing:* Because of all of the above, a technological decision in the public domain should never be framed entirely as a technical, or propositional, problem. Nonscientific preferences will always enter the decision to a greater or lesser degree.

BIBLIOGRAPHY

Ainley, P., and H. Rainbird, eds. 1999. *Apprenticeship: Towards a New Paradigm of Learning*. London: Kogan Page.

Arksey, Hilary. 1998. *RSI and the Experts: The Construction of Medical Knowledge*. London: UCL Press.

Bijker, Wiebe, E. 1995. *Of Bicycles, Bakelites, and Bulbs: Toward a Theory of Sociotechnical Change*. Cambridge, Mass.: MIT Press.

Bijker, W., T. Hughes, and T. Pinch, eds. 1987. *The Social Construction of Technological Systems*. Cambridge, Mass.: MIT Press.

Bloor, David. 1973. "Wittgenstein and Mannheim on the Sociology of Mathematics." *Studies in the History and Philosophy of Science* 4:173–91.

Bloor, David. 1983. *Wittgenstein: A Social Theory of Knowledge*. London: Macmillan.

Brannigan, G. 1981. *The Social Basis of Scientific Discoveries*. New York: Cambridge University Press.

Cabinet Office. 2000. *Code of Conduct for Written Consultations*. London: Cabinet Office. Available online at: http://archive.cabinetoffice.gov.uk/servicefirst/2000/consult/code/consultationcode.htm (accessed 13 December 2006).

Cabinet Office. 2005. *Code of Practice on Consultation*. London: Cabinet Office. Available online at: http://www.cabinetoffice.gov.uk/regulation/consultation/code/index.asp (accessed 13 December 2006).

Carolan, Michael, S. 2006. "Sustainable Agriculture, Science and the Co-production of 'Expert' Knowledge: The Value of Interactional Expertise." *Local Environment* 11:421–31.

Champod, Christophe, and Ian W. Evett. 2001. "A Probabilistic Approach to Fingerprint Evidence." *Journal of Forensic Identification* 51:101–22.

Collins, Harry. 1974. "The TEA Set: Tacit Knowledge and Scientific Networks." *Science Studies* 4:165–86.

Collins, Harry. 1987. "Certainty and the Public Understanding of Science: Science on Television." *Social Studies of Science* 17:689–713.

Collins, Harry. 1988. "Public Experiments and Displays of Virtuosity: The Core-Set Revisited." *Social Studies of Science* 18:725–48.

Collins, Harry. 1990. *Artificial Experts: Social Knowledge and Intelligent Machines*. Cambridge, Mass.: MIT Press.

Collins, Harry. 1992. *Changing Order: Replication and Induction in Scientific Practice*.

Chicago: University of Chicago Press. First ed., Beverley Hills, Calif., and London: Sage, 1985.

Collins, Harry. 1996a. "Embedded or Embodied: Hubert Dreyfus's *What Computers Still Can't Do." Artificial Intelligence* 80, no. 1: 99–117.

Collins, Harry. 1996b. "Interaction Without Society? What Avatars Can't Do." In *Internet Dreams,* edited by M. Stefik, 317–26. Cambridge, Mass.: MIT Press.

Collins, Harry. 1998. "The Meaning of Data: Open and Closed Evidential Cultures in the Search for Gravitational Waves." *American Journal of Sociology* 104, no. 2: 293–337.

Collins, Harry. 1999. "Tantalus and the Aliens: Publications, Audiences and the Search for Gravitational Waves." *Social Studies of Science* 29, no. 2: 163–97.

Collins, Harry. 2000. "Four Kinds of Knowledge, Two (or maybe Three) Kinds of Embodiment, and the Question of Artificial Intelligence." In *Heidegger, Coping, and Cognitive Science: Essays in Honor of Hubert L. Dreyfus,* vol. 2, edited by Jeff Malpas and Mark A. Wrathall, 179–95. Cambridge, Mass.: MIT Press.

Collins, Harry. 2001a. "What is Tacit Knowledge?" In *The Practice Turn in Contemporary Theory,* edited by Theodore R. Schatzki, Karin Knorr-Cetina and Eike von Savigny, 107–19. London: Routledge.

Collins, Harry. 2001b. "Tacit Knowledge, Trust, and the Q of Sapphire." *Social Studies of Science* 31, no. 1: 71–85.

Collins, Harry. 2004a. *Gravity's Shadow: The Search for Gravitational Waves.* Chicago: University of Chicago Press.

Collins, Harry. 2004b. "Interactional Expertise as a Third Kind of Knowledge." *Phenomenology and the Cognitive Sciences* 3, no. 2: 125–43.

Collins, Harry. 2004c. "The Trouble with Madeleine." *Phenomenology and the Cognitive Sciences* 3, no. 2: 165–70.

Collins, Harry. 2007. "Bicycling on the Moon: Collective Tacit Knowledge and Somatic-limit Tacit Knowledge." *Organization Studies* 0:000–000.

Collins, Harry. 2008, forthcoming. "Mathematical Understanding and the Physical Sciences." In *Case Studies of Expertise and Experience,* edited by Harry Collins, a special issue of *Studies in History and Philosophy of Science* 39, no. 1 (March).

Collins, Harry, and Robert Evans. 2002. "The Third Wave of Science Studies: Studies of Expertise and Experience." *Social Studies of Sciences* 32, no. 2: 235–96. Reprinted in Selinger and Crease 2006, 39–110.

Collins, Harry, Robert Evans, Rodrigo Ribeiro, and Martin Hall. 2006. "Experiments with Interactional Expertise." *Studies in History and Philosophy of Science* 37, A/4 (December: 656–74.

Collins, Harry, Rodney Green, and Bob Draper. 1985. "Where's the Expertise? Expert Systems as a Medium of Knowledge Transfer." In *Expert Systems 85,* edited by M. J. Merry, 323–34. Cambridge: Cambridge University Press.

Collins, Harry, and Martin Kusch. 1998. *The Shape of Actions: What Humans and Machines Can Do.* Cambridge, Mass.: MIT Press.

Collins, Harry, and Trevor Pinch. 1979. "The Construction of the Paranormal: Nothing Unscientific is Happening." In *Sociological Review Monograph,* no. 27: *On the Margins of Science: The Social Construction of Rejected Knowledge,* edited by Roy Wallis, 237–70. Keele: Keele University Press.

Collins, Harry, and Trevor Pinch. 1993/1998. *The Golem: What You Should Know About Science.* Cambridge and New York: Cambridge University Press. Second ed., Cambridge: Canto, 1998.

Collins, Harry, and Trevor Pinch. 1998. *The Golem at Large: What You Should Know About Technology.* Cambridge: Cambridge University Press.

Collins, Harry, and Trevor Pinch. 2005. *Dr Golem: What You Should Know about Medical Science.* Chicago: University of Chicago Press.

Collins, Harry, and Gary Sanders. 2008, forthcoming. "'They Give You the Keys and Say 'Drive It!' Managers, Referred Expertise, and Other Expertises." In *Case Studies of Expertise and Experience,* edited by Harry Collins, a special issue of *Studies in History and Philosophy of Science* 39, no. 1 (March).

Collins, Harry, and Steven Yearley. 1992. "Epistemological Chicken." In *Science as Practice and Culture,* edited by A. Pickering, 301–26. Chicago: University of Chicago Press.

Coy, M. W., ed. 1989. *Apprenticeship: From Theory to Method and Back Again.* Albany: State University of New York Press.

Crease, Robert, P. 2003. "Inquiry and Performance: Analogies and Identities Between the Arts and the Sciences." *Interdisciplinary Science Reviews* 28:267–72.

Dawkins, Richard. 1999. *Unweaving the Rainbow: Science, Delusion and the Appetite for Wonder.* London: Penguin.

Dear, Peter. 1995. *Discipline and Experience: The Mathematical Way in the Scientific Revolution.* Chicago: University of Chicago Press.

Dreyfus, Hubert L. 1967. "Why Computers Must Have Bodies in Order to be Intelligent." *The Review of Metaphysics* 21, no. 1: 13–32.

Dreyfus, Hubert L. 1972. *What Computers Can't Do.* New York: Harper and Row.

Dreyfus, Hubert L. 1992. *What Computers Still Can't Do.* Cambridge, Mass.: MIT Press.

Dreyfus, Hubert L., and Stuart E. Dreyfus. 1986. *Mind Over Machine: The Power of Human Intuition and Expertise in the Era of the Computer.* New York: Free Press.

Epstein, Steven. 1995. "The Construction of Lay Expertise: AIDS Activism and the Forging of Credibility in the Reform of Clinical Trials." *Science Technology and Human Values* 20:408–37.

Epstein, Steven 1996. *Impure Science: AIDS, Activism and the Politics of Knowledge.* Berkeley and Los Angeles: University of California Press.

European Commission. 2001. *European Governance: A White Paper. (COM [2001] 428 final.)* Brussels: European Commission.

European Commission. 2002. *The Collection and Use of Expertise by the Commission: Principles and Guidelines. Improving the Knowledge Base for Better Policies. (COM [2002] 713 final.)* Brussels: European Commission.

Evans, Robert. 1999. *Macroeconomic Forecasting: A Sociological Appraisal.* London: Routledge.

Fleck, Ludwik. 1979. *Genesis and Development of a Scientific Fact.* Chicago: University of Chicago Press. First published in German in 1935.

Fulton, Lord. 1968. *The Civil Service,* vol. 2: *Report of a Management Consultancy Group. Evidence submitted to the Committee under the Chairmanship of Lord Fulton, 1966–1968.* London: Her Majesty's Stationery Office.

Ginzburg, Carlo. 1989. "Morelli, Freud and Sherlock Holmes: Clues and Scientific Method." *History Workshop Journal* 9:5–36.

Goldman, Alvin I. 2001. "Experts: Which Ones Should You Trust?" *Philosophy and Phenomenological Research* 63, no. 1: 85–110. Reprinted in Selinger and Crease 2006, 14–38.

Goodman, Nelson. 1969. *Languages of Art.* London: Oxford University Press.

Gough, C. 2000. "Science and the Stradivarius." *Physics World* 13, no. 4: 2–33.

Gross, Paul, and Norman Levitt. 1994. *Higher Supersition: The Academic Left and its Quarrels with Science.* Baltimore and London: John Hopkins University Press.

Gross, Paul, Norman Levitt, and M. W. Lewis. 1996. *The Flight From Science and Reason.* New York: New York Academy of Sciences.

Guston, David H. 1999. "Evaluating the First U.S. Consensus Conference: The Impact of the Citizens' Panel on Telecommunications and the Future of Democracy." *Science, Technology and Human Values* 24, no. 4: 451–82.

Halfpenny, Peter. 1982. *Positivism and Sociology: Explaining Social Life.* London: George Allen and Unwin.

Harvey, B. 1981. "Plausibility and the Evaluation of Knowledge: A Case Study in Experimental Quantum Mechanics." *Social Studies of Science* 11:95–130.

Hennessy, Peter 1989. *Whitehall.* London: Martin Secker and Warburg Ltd.

House of Lords. 2000. *Science and Society: Science and Technology Select Committee, Third Report.* London: HMSO. Also available at: http://www.parliament.the-stationery-office.co.uk/pa/ld199900/ldselect/ldsctech/38/3801.htm (accessed 23 January 2002) and at http://www.economics.unimelb.edu.au/research/workingpapers/wp97_99/715.html (accessed 15 June 2003).

Ihde, Don. 1997. "Why Not Science Critics?" *International Studies in Philosophy* 29:45–54. Reprinted in Selinger and Crease 2006, 39–403.

Irwin, Alan, and Brian Wynne, eds. 1996. *Misunderstanding Science? The Public Reconstruction of Science and Technology.* Cambridge and New York: Cambridge University Press.

Jackson, F. 1986. "What Mary Didn't Know." *Journal of Philosophy* 83:291–95.

Koertge, Noretta, ed. 2000. *A House Built on Sand: Exposing Postmodernist Myths About Science.* Oxford: Oxford University Press.

Kosinski, Jerzy 1971. *Being There.* Orlando: Harcourt Brace Jovanovich, Inc.

Kuhn, Thomas S. 1962. *The Structure of Scientific Revolutions.* Chicago: University of Chicago Press.

Kusch, Martin. 2002. *Knowledge by Agreement: The Programme of Communitarian Epistemology.* Oxford: Oxford University Press.

Kusch, Martin. 2007. "Towards a Political Philosophy of Risk: Experts and Publics in Deliberative Democracy." In *Risk: Philosophical Perspectives,* edited by Tim Lewens. London: Routledge.

Ladd, Paddy. 2003. *Understanding Deaf Culture: In Search of Deafhood.* Clevedon: Multilingual Matters, Ltd.

Latour, Bruno, and S. Woolgar. 1979. *Laboratory Life: The Social Construction of Scientific Facts.* London and Beverly Hills: Sage.

Lave, Jean 1988. *Cognition in Practice.* Cambridge: Cambridge University Press.

Lave, Jean, and Etienne Wenger. 1991. *Situated Learning: Legitimate Peripheral Participation.* Cambridge: Cambridge University Press.

Lawless, Edward W. 1977. *Technology and Social Shock.* New Brunswick, N.J.: Rutgers University Press.

Lynch, Michael, and Simon Cole. 2005. "Science and Technology Studies on Trial." *Social Studies of Science* 35, no. 2: 269–311.

MacKenzie, Donald. 1998. "The Certainty Trough." In *Exploring Expertise: Issues and Perspectives,* edited by R. Williams, W. Faulkner, and J. Fleck, 325–29. Basingstoke: Macmillan.

Maurer, D. W. 1940. *The Big Con: The Story of the Confidence Man and the Confidence Game.* New York: Bobs Merrill.

Muir, Frank. 1997. *A Kentish Lad*. Reading: Corgi Books.

Murcott, Anne. 1999. "Not Science but PR: GM Food and the Makings of a Considered Sociology." *Sociological Research Online* 4:3.

Murcott, Anne. 2001. "Public Beliefs about GM Foods: More on the Makings of a Considered Sociology." *Medical Anthropology Quarterly* 15, no. 1: 1–11.

Naftulin, Donald H., John E. Ware Jr, and Frank A. Donnelly. 1973. "The Doctor Fox Lecture: A Paradigm of Educational Seduction." *Journal of Medical Education* 48:630–35.

Nonaka, Ikujiro, and Hirotaka Takeuchi. 1995. *The Knowledge-Creating Company: How Japanese Companies Create the Dynamics of Innovation*. Oxford: Oxford University Press.

Office of Science and Technology and the Wellcome Trust. 2000. *Science and the Public: A Review of Science Communication and Public Attitudes to Science in Britain*, London: Wellcome Trust. Available at: http://www.wellcome.ac.uk/en/1/pinpubactconpub.html (accessed 17 December 2003).

Pamplin, B. R., and H. M. Collins. 1975. "Spoon Bending: An Experimental Approach." *Nature* 257:8 (4 September).

Peterson, J. C., and G. E. Markle. 1979. "Politics and Science in the Laetrile Controversy." *Social Studies of Science* 9, no. 2: 139–66.

Pinch, T., H. M. Collins, and L. Carbone. 1996. "Inside Knowledge: Second Order Measures of Skill." *Sociological Review* 44, no. 2: 163–86.

Polanyi, Michael. 1958. *Personal Knowledge*. London: Routledge and Kegan Paul.

Polanyi, Michael. 1962. "The Republic of Science, Its Political and Economic Theory." *Minerva* 1:54–73.

The Politics of GM Food: Risk, Science and Public Trust. 1999. ESRC Special Briefing No. 5, October 1999. Swindon: Economic and Social Research Council. Available at http://www.sussex.ac.uk/Units/gec/gecko/gm-brief.htm (accessed 12 December 2006).

Popper, Karl R. 1957. *The Poverty of Historicism*. London: Routledge and Kegan Paul.

Pye, D. 1968. *The Nature and Art of Workmanship*. Cambridge: Cambridge University Press.

Ribeiro, R. and Collins, H. M. 2007. "The Bread-making Machine, Tacit Knowledge and the Theory of Action Morphicity." *Organization Studies* 0/00: 000–00.

Sacks, Oliver. 1985. *The Man Who Mistook his Wife for a Hat*. London: Duckworth.

Sacks, Oliver 1989. *Seeing Voices: A Journey into the World of the Deaf*. Berkeley and Los Angeles: University of California Press.

Sclove, Dick. 1997. "Telecommunications and the Future of Democracy: Preliminary Report on the First U.S. Citizens' Panel." *Loka Alert* 4:3 (April). Available at http://www.loka.org/alerts/loka.4.3.htm (accessed 17 December 2003).

Scott, David, and Alexei Leonov. 2004. *Two Sides of the Moon: Our Story of the Cold War Space Race*. London: Simon and Schuster.

Selinger, Evan. 2003. "The Necessity of Embodiment: The Dreyfus-Collins Debate." *Philosophy Today* 47, no. 3: 266–79.

Selinger, Evan, Hubert Dreyfus, and Harry Collins. 2008, forthcoming. "Interactional Expertise and Embodiment." In *Case Studies of Expertise and Experience*, edited by Harry Collins, a special issue of *Studies in History and Philosophy of Science*, 39, no. 1 (March).

Selinger, Evan, and Robert Crease, eds. 2006. *The Philosophy of Expertise*. New York: Columbia University Press.

Selinger, Evan, and Tom Mix. 2004. "On Interactional Expertise: Pragmatic and Onto-
logical Considerations." *Phenomenology and the Cognitive Sciences* 3, no. 2: 145–63.
Reprinted in Selinger and Crease 2006, 302–21.

Shapin, Steven. 1979. "The Politics of Observation: Cerebral Anatomy and Social
Interests in the Edinburgh Phrenology Disputes." In *On the Margins of Science: The
Social Construction of Rejected Knowledge*, Sociological Review Monograph 27, edited
by R. Wallis, 139–78. Keele: Keele University Press.

Shapin, Steven, and Simon Schaffer. 1987. *Leviathan and the Air Pump: Hobbes, Boyle
and the Experimental Life*. Princeton: Princeton University Press.

Sokal, Alan. 1996. "Transgressing the Boundaries: Towards a Transformative Hermeneutics of Quantum Gravity." *Social Text* 46/47: 217–52.

Speers, T., and J. Lewis. 2003. "MMR and the Media: Misleading Reporting?" *Nature
Reviews, Immunology* 3, no. 11: 913–18.

Speers, T., and J. Lewis. 2004. "Jabbing the Scientists: Media Coverage of the MMR
Vaccine in 2002." *Communication and Medicine* 1, no. 2: 171–82.

Suchman, Lucy A. 1987. *Plans and Situated Action: The Problem of Human-machine Interaction*. Cambridge: Cambridge University Press.

Surowiecki, James. 2004. *The Wisdom of Crowds: Why the Many are Smarter than the Few*.
London: Little, Brown.

Tart, Charles T. 1972. "States of Consciousness and State-Specific Sciences." *Science*
176:1203–10.

Thorpe, Charles. 2002. "Disciplining Experts: Scientific Authority and Liberal Democracy in the Oppenheimer Case." *Social Studies of Science* 34, no. 4: 525–62.

Thorpe, Charles, and Steven Shapin. 2000. "Who Was J. Robert Oppenheimer? Cha-
risma and Complex Organization." *Social Studies of Science* 30, no. 4: 545–90.

Turing, A. M. 1950. "Computing Machinery and Intelligence." *Mind* 59:433–60.

Von-Hippel, Eric. 1988. *The Sources of Innovation*. New York: Oxford University Press.

Weizenbaum, J. 1976. *Computer Power and Human Reason: From Judgment to Calculation*.
San Francisco: W. H. Freeman.

Welsh, Ian 2000. *Mobilising Modernity: The Nuclear Moment*. London: Routledge.

Winch, Peter G. 1958. *The Idea of a Social Science*. London: Routledge and Kegan Paul.

Winograd, T., and F. Flores. 1986. *Understanding Computers and Cognition: A New Foun-
dation for Design*. New Jersey: Ablex.

Wittgenstein, Ludwig. 1953. *Philosophical Investigations*. Oxford: Blackwell.

Wolpert, Lewis. 1992. *The Unnatural Nature of Science*. London: Faber and Faber.

Wolpert, Lewis. 1994. "Review of *The Golem: What Everyone Should Know About Science*." *Public Understanding of Science* 3 :323–37.

Wynne, Brian. 1989. "Sheep Farming after Chernobyl: A Case Study in Communicat-
ing Scientific Information." *Environmental Magazine* 31, no. 2: 33–39.

Wynne, Brian. 1992. "Public Understanding of Science Research: New Horizon or
Hall of Mirrors?" *Public Understanding of Science* 1, no. 1: 37–43.

Wynne, Brian. 1993. "Public Uptake of Science: A Case for Institutional Reflexivity."
Public Understanding of Science 2, no. 4: 321–37.

Wynne, Brian. 1996a. "May the Sheep Safely Graze? A Reflexive View of the Expert-
Lay Knowledge Divide." In *Risk, Environment and Modernity: Towards a New Ecology*,
edited by S. Lash, B. Szerszynski, and B. Wynne, 27–83. London: Sage.

Wynne, Brian. 1996b. "Misunderstood Misunderstandings: Social Identities and
Public Uptake of Science." In *Misunderstanding Science? The Public Reconstruction of*

Science and Technology, edited by Alan Irwin and Brian Wynne, 19–46. Cambridge: Cambridge University Press.

Wynne, Brian. 2003. "Seasick on the Third Wave? Subverting the Hegemony of Pro-positionalism." *Social Studies of Science* 33, no. 3: 401–17.

von Wright, G. H. 1971. *Explanation and Understanding.* London: Routledge and Kegan Paul.

INDEX